I0037653

Nanoporous Materials and Their Applications

Nanoporous Materials and Their Applications

Special Issue Editors

Enrique Rodríguez-Castellón
Sibele Pergher

MDPI • Basel • Beijing • Wuhan • Barcelona • Belgrade

MDPI

Special Issue Editors
Enrique Rodríguez-Castellón
University of Málaga
Spain

Sibele Pergher
Universidade Federal do Rio Grande do Norte
Brazil

Editorial Office
MDPI
St. Alban-Anlage 66
4052 Basel, Switzerland

This is a reprint of articles from the Special Issue published online in the open access journal *Applied Sciences* (ISSN 2076-3417) from 2018 to 2019 (available at: https://www.mdpi.com/journal/applsci/special_issues/Nanoporous_Materials)

For citation purposes, cite each article independently as indicated on the article page online and as indicated below:

LastName, A.A.; LastName, B.B.; LastName, C.C. Article Title. *Journal Name* **Year**, *Article Number, Page Range.*

ISBN 978-3-03897-968-5 (Pbk)
ISBN 978-3-03897-969-2 (PDF)

© 2019 by the authors. Articles in this book are Open Access and distributed under the Creative Commons Attribution (CC BY) license, which allows users to download, copy and build upon published articles, as long as the author and publisher are properly credited, which ensures maximum dissemination and a wider impact of our publications.

The book as a whole is distributed by MDPI under the terms and conditions of the Creative Commons license CC BY-NC-ND.

Contents

About the Special Issue Editors

Enrique Rodríguez-Castellón has a degee in chemistry (Universidad Autónoma de Madrid), a master's degree in chemistry (University of Puerto Rico), and a doctorate in chemistry (Universidad de Málaga). He is a professor of inorganic chemistry at the Universidad de Málaga. He is the president of the Inorganic Chemistry Division of the Spanish Royal Society of Chemistry. He has published more than 475 papers on materials science and catalysis. He holds six patents and has completed more than 45 research projects and contracts. He has an H index of 52 and more than 11,500 citations. He was recently awarded Professor Honoris Causa of the Federal University of Ceará (Brazil).

Sibele Pergher, Ph.D., has a degree in chemical engineering (UFRGS, 1990), a master's degree in chemical engineering (UEM, 1993), and a doctorate in chemistry (UPV/ITQ- Spain, 1997). She worked at UFRGS (1998–2001) and at URI - Erechim (2001–2010). She is currently a professor and researcher (since 2010) at UFRN. She is also a director of the Brazilian Society of Catalysis—SBCat, part of the Synthesis Commission of IZA, and represents Brazil in FISOCAT and IACS. She is the coordinator and founder of LABPEMOL. She works mainly on the synthesis of catalysts, zeolites, clays, mesoporous materials, lamellar, and pillared and delaminate materials in adsorption and catalysis processes. She has published more than 150 papers, holds 20 patents, and completed 500 work on congress. She has also made a great contribution to the training of academic students, completing more than 150 orientations.

applied
sciences

MDPI

Editorial

Nanoporous Materials and Their Applications

Sibele B. C. Pergher [1,*] and Enrique Rodríguez-Castellón [2]

[1] Departamento de Química, Universidade Federal do Rio Grande do Norte, Natal Caixa postal 1524, Brazil
[2] Department of Inorganic Chemistry, Faculty of Science, University of Malaga, 29071 Malaga, Spain;
     castellon@uma.es
*   Correspondence: sibelepergher@gmail.com; Tel.: +55-84-9941-35418

Received: 12 March 2019; Accepted: 18 March 2019; Published: 29 March 2019

Investigations into nanoporous materials and their applications continue to afford a wealth of novel materials and new applications. In fact, the ongoing quest for nanoporous materials with novel properties has led to many new materials and new applications for known materials.

This Special Issue is associated with the most recent advances in nanoporous material synthesis, as well as its applications.

The 12 articles comprising this Special Issue can be considered a representative selection of the current research on this topic, reflecting the diversity of nanoporous materials and their applications.

For example, Schwanke and Pergher [1] provide a review of nanoporous materials with MWW topology. They cover aspects of the synthesis of the MWW precursor and the tridimensional zeolite MCM-22, as well as their physicochemical properties, such as the Si/Al molar ratio, acidity, and morphology. In addition, this paper discusses the use of directing agents (SDAs) to obtain the different MWW-type materials reported thus far. The traditional post-synthesis modifications to obtain MWW-type materials with hierarchical architectures, such as expanded, swelling, pillaring, and delaminating structures, are shown together with recent routes to obtain materials with more open structures. New routes for the direct synthesis of MWW-type materials with a hierarchical pore architecture are also covered.

Silva et al. [2] study hierarchical materials by a method of opening mesopores in a microporous zeolite structure. They created mesopores in the Ultrastable USY zeolite (Si/Al = 15) using alkaline treatment (NaOH) in the presence of cetyltrimethylammonium bromide surfactant, followed by hydrothermal treatment. The effects of the different concentrations of NaOH and the surfactant on the textural, chemical, and morphological characteristics of the modified zeolites are evaluated.

Also in the area of zeolite synthesis, Vinaches et al. [3] propose an alternative method for the introduction of aluminum into the STW zeolitic framework. This zeolite was synthesized in a pure silica form, and an aluminum source was added by in situ generated seeds. Characterization techniques, such as XRD and MAS NMR of 29Si and 27Al, were used to conclude that the aluminum was effectively introduced into the framework. The materials were tested as catalysts on the dehydration of ethanol, and they proved be selective to ethylene and diethyl ether, confirming the presence of acidic sites.

Another approach was analyzed by the group of Pereira et al. [4], who study the synthesis of zeolites from two metakaolins, one derived from the white kaolin and the other derived from the red kaolin, found in a deposit in the city of São Simão (Brazil). The A zeolite obtained was applied as an adsorbent to remove methylene blue, safranine, and malachite green from aqueous solutions.

Another applications approach was proposed by Zhang et al. [5], who compared two glycerol/ ZSM-5 zeolite systems with different amount of residual gas by performing a series of experiments.

Besides zeolite materials, there exists one kind of micro and mesoporos materials built by pillarization of lamellar materials. On this subject, the paper of Jalil et al. [6] is very interesting. Jalil et al. synthesized, characterized, and evaluated three silica pillared clays as possible adsorbents of ciprofloxacin (CPX) and tetracycline (TC) from alkaline aqueous media.

Another natural raw material was used in the separations process. Autie-Pérez et al. [7] study a raw porous volcanic glass from Cuba as an adsorbent for Cu^{2+} removal from dyes after activation with an acid solution. After Cu^{2+} adsorption, its capacity to separate n-paraffins from a mixture by inverse gas chromatography (IGC) was also evaluated. They showed that natural volcanic glass can be used in both heavy metal removal and paraffin separation for industrial purposes.

Mesoporous materials are very interesting because of their great accessibility to bulk molecules. On this subject, we have the study of Fernandes et al. [8], which analyzes the influence of Synthesis Parameters in Obtaining KIT-6 Mesoporous Materials, and the study of Busatta et al. [9], which examines the ethylene oligomerization reactions catalyzed by nickel-β-diimine complexes immobilized on β-zeolite, [Si]-MCM-41 and [Si,Al]-MCM-41, modified with an ionic liquid. They showed different selectivities depending on whether the material used zeolite (microporous) or MCM41 (mesoporous) materials. Also on this subject, Padula et al. [10] synthesized Mesoporous Niobium Oxyhydroxide Catalysts for Cyclohexene Epoxidation Reactions. These mesoporous catalysts were synthesized from the precursor NbCl5 and surfactant CTAB (cetyltrimethylammonium bromide) using different synthesis routes, in order to obtain materials with different properties, which are capable of promoting the epoxidation of cyclohexene. Catalytic studies have shown that mild reaction conditions promote high conversion.

Another interesting type of porous material are the MOFs, or Metal Organic Frameworks. Fuentes-Fernandez et al. [11] study the confined porous environment of MOFs as a system for studying reaction mechanisms. As an example of an important reaction, they study the dissociation of water—which plays a critical role in biology, chemistry, and materials science—in MOFs and show how the knowledge of the structure in this confined environment allows for an unprecedented level of understanding and control. Their results show that precise control of reactions within nano-porous materials is possible, opening the way for advances in fields ranging from catalysis to electrochemistry and sensors. Wu et al. [12] present an experimental investigation into the third-order nonlinearity of conventional crystalline (c-Si) and porous (p-Si) silicon with a Z-scan technique at 800-nm and 2.4-μm wavelengths.

Finally, I wish to express my gratitude to all the authors for their contributions to this Special Issue. I would also like to thank the reviewers for their kind, essential advice and suggestions. The contributions of the editorial, as well as the publishing, staff at Applied Science to this Special Issue are also highly appreciated. I hope readers from different research fields will enjoy this Open Access Special Issue and find a basis for further work in the exciting field of nanoporous materials.

References

1. Schwanke, A.; Pergher, S. Lamellar MWW-Type Zeolites: Toward Elegant Nanoporous Materials. *Appl. Sci.* **2018**, *8*, 1636. [CrossRef]
2. Silva, J.F.; Ferracine, E.D.; Cardoso, D. Effects of Different Variables on the Formation of Mesopores in Y Zeolite by the Action of CTA+ Surfactant. *Appl. Sci.* **2018**, *8*, 1299. [CrossRef]
3. Vinaches, P.; Rojas, A.; De Alencar, A.E.V.; Rodríguez-Castellón, E.; Braga, T.P.; Pergher, S.B.C. Introduction of Al into the HPM-1 Framework by In Situ Generated Seeds as an Alternative Methodology. *Appl. Sci.* **2018**, *8*, 1634. [CrossRef]
4. Pereira, P.M.; Ferreira, B.F.; Oliveira, N.P.; Nassar, E.J.; Ciuffi, K.J.; Vicente, M.A.; Trujillano, R.; Rives, V.; Gil, A.; Korili, S.; et al. Synthesis of Zeolite A from Metakaolin and Its Application in the Adsorption of Cationic Dyes. *Appl. Sci.* **2018**, *8*, 608. [CrossRef]
5. Zhang, Y.; Luo, R.; Zhou, Q.; Chen, X.; Dou, Y. Effect of Degassing on the Stability and Reversibility of Glycerol/ZSM-5 Zeolite System. *Appl. Sci.* **2018**, *8*, 1065. [CrossRef]
6. Roca Jalil, M.E.; Toschi, F.; Baschini, M.; Sapag, K. Silica Pillared Montmorillonites as Possible Adsorbents of Antibiotics from Water Media. *Appl. Sci.* **2018**, *8*, 1403. [CrossRef]

7. Autie-Pérez, M.; Infantes-Molina, A.; Cecilia, J.A.; Labadie-Suárez, J.M.; Rodríguez-Castellón, E. Separation of Light Liquid Paraffin C_5–C_9 with Cuban Volcanic Glass Previously Used in Copper Elimination from Water Solutions. *Appl. Sci.* **2018**, *8*, 295. [CrossRef]
8. Fernandes, F.R.D.; Pinto, F.G.H.S.; Lima, E.L.F.; Souza, L.D.; Caldeira, V.P.S.; Santos, A.G.D. Influence of Synthesis Parameters in Obtaining KIT-6 Mesoporous Material. *Appl. Sci.* **2018**, *8*, 725. [CrossRef]
9. Busatta, C.A.; Mignoni, M.L.; De Souza, R.F.; Bernardo-Gusmão, K. Nickel Complexes Immobilized in Modified Ionic Liquids Anchored in Structured Materials for Ethylene Oligomerization. *Appl. Sci.* **2018**, *8*, 717. [CrossRef]
10. Padula, I.D.; Chagas, P.; Furst, C.G.; Oliveira, L.C.A. Mesoporous Niobium Oxyhydroxide Catalysts for Cyclohexene Epoxidation Reactions. *Appl. Sci.* **2018**, *8*, 881. [CrossRef]
11. Fuentes-Fernandez, E.M.A.; Jensen, S.; Tan, K.; Zuluaga, S.; Wang, H.; Li, J.; Thonhauser, T.; Chabal, Y.J. Controlling Chemical Reactions in Confined Environments: Water Dissociation in MOF-74. *Appl. Sci.* **2018**, *8*, 270. [CrossRef]
12. Wu, R.; Collins, J.; Canham, L.T.; Kaplan, A. The Influence of Quantum Confinement on Third-Order Nonlinearities in Porous Silicon Thin Films. *Appl. Sci.* **2018**, *8*, 1810. [CrossRef]

© 2019 by the authors. Licensee MDPI, Basel, Switzerland. This article is an open access article distributed under the terms and conditions of the Creative Commons Attribution (CC BY) license (http://creativecommons.org/licenses/by/4.0/).

applied
sciences

MDPI

Review

Lamellar MWW-Type Zeolites: Toward Elegant Nanoporous Materials

Anderson Schwanke [1,*] and Sibele Pergher [2]

[1]   Instituto de Química, Laboratório de Reatividade e Catálise, Universidade Federal do Rio Grande do Sul,
      Porto Alegre 91540-000, RS, Brazil
[2]   Instituto de Química, Laboratório de Peneiras Moleculares (LABPEMOL), Universidade Federal do Rio
      Grande do Norte, Natal 59078-970, RN, Brazil; sibelepergher@gmail.com
*   Correspondence: anderson-js@live.com; Tel.: +55-54-98129-3396

Received: 11 July 2018; Accepted: 7 August 2018; Published: 13 September 2018

Featured Application: This work is a compilation of different strategies to obtain lamellar zeolitic materials with a hierarchical structure of pores. The aim of this work is to offer a greater dissemination of MWW-type lamellar zeolites to demonstrate the most recent strategies for obtaining materials with different pore architectures and providing promising applications in catalysis, adsorption, and separation.

Abstract: This article provides an overview of nanoporous materials with MWW (Mobil twenty two) topology. It covers aspects of the synthesis of the MWW precursor and the tridimensional zeolite MCM-22 (Mobil Composition of Matter number 22) as well as their physicochemical properties, such as the Si/Al molar ratio, acidity, and morphology. In addition, it discusses the use of directing agents (SDAs) to obtain the different MWW-type materials reported so far. The traditional post-synthesis modifications to obtain MWW-type materials with hierarchical architectures, such as expanded, swelling, pillaring, and delaminating structures, are shown together with recent routes to obtain materials with more open structures. New routes for the direct synthesis of MWW-type materials with hierarchical pore architecture are also covered.

Keywords: zeolite; MWW; MCM-22; hierarchical zeolite; lamellar zeolite; layered zeolite; two-dimensional zeolites; swelling; pillaring; delaminating

---

1. Introduction

Zeolites are a class of crystalline materials formed by a skeleton based on tetrahedral silicon and aluminum (and others, such as P, Ge, Ga, B, S, and Fe), which form microporous (<2 nm) channels and cavities. Due to their microporous structure, these materials are extremely versatile and are widely used as adsorbents, ion exchangers, detergents, and catalysts [1–3]. However, it is in the catalysis field that zeolites play an essential role in refining, processing, and organic synthesis for fine chemistry. In fact, zeolites make up more than 40% of the solid catalysts used in the chemical industry [4].

In the last two decades, two-dimensional lamellar zeolitic precursors (LZPs) have been found for some types of zeolites. These LZPs show the same basic structure as the tridimensional form with separated lamellae approximately 1 to 2 nm thick along one direction, and these precursors condense topotactically, producing three-dimensional structures. According to the International Zeolite Association (IZA), there are more than 200 framework topologies, and less than 10% of these structures have an LZP or exist in a two-dimensional form [5]. MWW, FER, NSI, OKO, RRO, CAS, CDO, PCR, RWR, and AFO are some examples of zeolite framework topologies that exist with a lamellar form. Readers can find the list of lamellar zeolites and their references in excellent reviews [6–10].

Among these framework topologies, LZP with MWW topology, known as (P)MCM-22 (MCM-22 precursor), is remarkably the most studied LZP. Moreover, its modification with postsynthetic procedures yields engineered materials with different pore architectures and lamellae organizations, such as hybrid organic-inorganic, pillared, misaligned, disordered, delaminated, and desilicated structures [11]. These modifications open avenues to obtaining elegant and designed solids with hierarchical pore structures that could facilitate reactants in reaching active sites to increase the conversion and yield of the desired products. Considering the importance and the versatility of this class of nanoporous materials, this work will focus on MWW-type zeolites, showing their general aspects of synthesis and recent advances.

2. The Precursor (P)MCM-22 and MCM-22 Zeolites

(P)MCM-22 was reported in 1990 by Mobil and is composed of individual lamellae with a thickness of 2.5 nm, with sinusoidal 10-ring channels (0.40 × 0.50 nm) and 12-ring hemicavities (connected to each other by double 6-rings with an aperture of ~0.3 nm) on the upper and lower surface of the lamella [12,13]. The precursor contains hexamethyleneimine (HMI) molecules used as a structure directing agent (SDA) occluded in the sinusoidal channels, as shown in Figure 1a. The interaction between lamellae occurs via hydrogen bonds between silanol groups on the surface and the HMI molecules also present between the MWW lamellae [8].

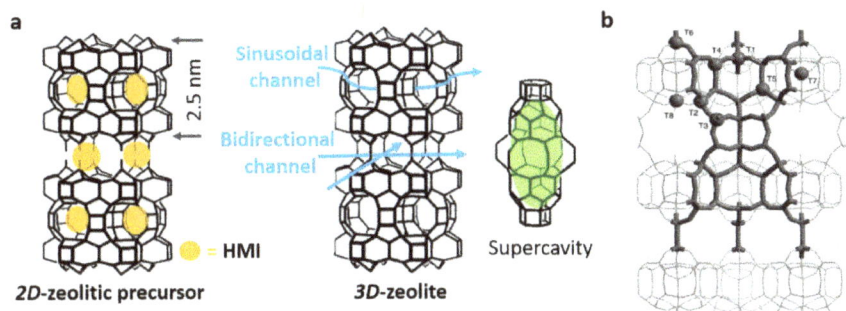

Figure 1. The two-dimensional (2D) zeolitic precursor and three-dimensional (3D) MWW (Mobil twenty two) zeolitic structure (**a**); its eight different tetrahedral sites (**b**); HMI (hexamethyleneimine). Adapted from Reference [14,15]. Copyright (1998), with the permission of the Royal Society of Chemistry, and Copyright (2006), with permission from Elsevier.

After calcination, the organic content is removed and the silanol groups between lamellae are condensed to form three-dimensional MCM-22 zeolite, as shown in Figure 1a. Consequently, an additional two-directional 10-ring (0.40 × 0.55 nm) channel, formed as the union of 12-ring hemicavities, generates internal super cages (free internal diameter of 0.71 nm and internal height of 1.82 nm), which are also connected to the aforementioned two-directional 10-ring channels [13]. In addition, it is possible to obtain a three-dimensional MCM-22 analog called MCM-49, which is obtained by direct crystallization by increasing the relative proportion of alkali in the gel composition [16]. Furthermore, another material called MCM-56 with a partial disorder of lamellar stacking is obtained when the reaction to form MCM-49 is stopped in the middle of the crystallization course [17,18].

The MCM-22 zeolite with high crystallinity can be obtained with Si/Al molar ratios between 15 and 70 (usually 20). However, ferrierite competing phases are found when the amount of aluminum increases (Si/Al = 9), as shown in the microscopic analysis of Figure 2 (image a, white arrows). In contrast, the decrease in aluminum in the synthesis gel (Si/Al > 70) leads to MFI (Mobil Five) competing phases [19]. It is also possible to obtain a pure silica zeolite MCM-22 analog by direct synthesis using mixtures of trimethyladamantylammonium (TMAda$^+$) and HMI as a second organic template, which is known as ITQ-1 [14].

The Si/Al ratio associated with the synthesis temperature and the static and dynamic (rotation of the autoclaves) conditions of the gel aging could interfere in the formation of MCM-22. It was reported that the use of 30 < Si/Al < 70 and temperatures above 150 °C results in the formation of ferrierite and/or mordenite competing phases. At temperatures below 150 °C with dynamic conditions, the MCM-22 phase is insignificant, while the formation of other phases increases with a decrease in the Si/Al ratio [20]. Other authors have reported that an MCM-22 formation with no other phase competitions was avoided using temperatures between 135 and 150 °C. Furthermore, dynamic conditions produce MCM-22 zeolite with good quality, while static conditions result in the nonsignificant formation of the desired phase or even the formation of pure ferrierite [20,21]. In addition, it was reported that the previous aging of the gel at 180 °C for 4–12 h and static conditions produced pure MCM-22 with a reduced crystallization time [22].

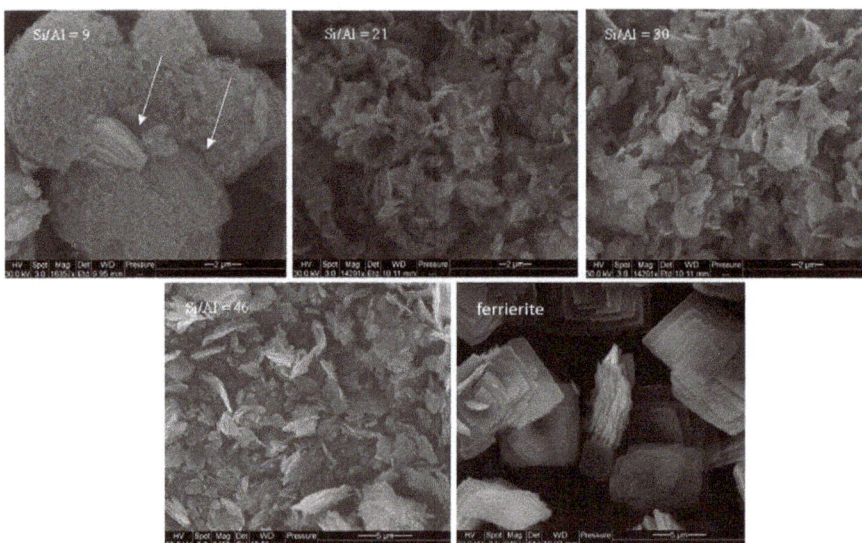

Figure 2. Morphologies of MCM-22 with different Si/Al molar ratios = 9, 21, 30, 46. The last image corresponds to pure ferrierite formed under static conditions. Adapted from Reference [19]. Copyright (2009), with permission from Elsevier.

MCM-22 presents distinct acid sites that reveal the homogeneity in acid strength. The microcalorimetry results showed a concentration of acid sites (for an MCM-22 with Si/Al = 16) of 1052 μmol·g^{-1}, which is modestly higher than the concentration of aluminum ions, 947 μmol·g^{-1}, suggesting that all aluminum ions produce acid sites either by producing an unbalanced charge structure that is balanced by the proton or active as Lewis acid sites. It was assumed that the aluminum ions in the zeolite structure do not act as Lewis acid sites because they are "protected" by nearby protonic centers. These aluminum ions may be located on the extra-framework where the structure is relaxed, acting as Lewis acid sites. The author of this study pointed out that the additional concentration of acid sites (105 μmol·g^{-1}) may be related to silanol groups located at the external surface [23].

Studies employing infrared spectroscopy with adsorbed pyridine have reported that for samples with Si/Al = 10, 14, and 30, most acid sites (50–70%) are located in the supercavities. The other sites are located in the sinusoidal channels (20–30%) or connected to the hexagonal prisms between supercavities, with values of 10% for Si/Al = 10 and 14 and 20% for Si/Al = 30 [24]. In addition, a study using density functional theory reported that the favorite placement sites of aluminum ions are the sites T1, T3, and T4, as shown in Figure 1b. The T2 site is presented as less favorite and the acidity of the T1 and T4 sites are equivalent and stronger than that of the T3 site, respectively [15].

Regarding the good performance of the MWW materials for benzene alkylation reactions, and despite the small size of the channel apertures, it is suggested that a significant number of cavities are open on the surface of the crystallites. It was assumed that the "cups" of the supercavities have a free diameter of 0.71 nm and the formation of cumene and ethylbenzene must occur in these cavities without any diffusional barrier. This hypothesis is supported when catalytic activity is significantly decreased by deactivation with 2,6-di-tert-butylpyridine, a large molecule that cannot enter in the channels of MCM-22. However, spectroscopic results confirm that benzene could easily enter the supercavities [23,25].

The hydrothermal crystallization and morphology of MCM-22 can be significantly altered by static or dynamic conditions. The dynamic condition minimizes the excessive aggregation of the crystals (see Figure 3a) when compared with static conditions, as shown in Figure 3b–g. Synthesis under dynamic conditions also induces the formation of zeolite with a higher crystallinity in a shorter time [17].

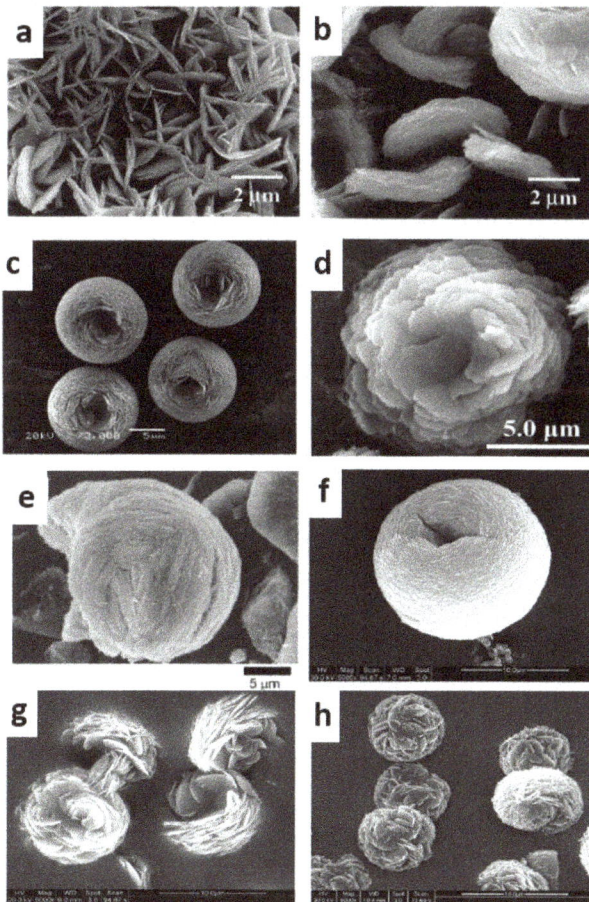

Figure 3. Morphologies of MCM-22 zeolites obtained under dynamic (**a**) and static conditions (**b–h**). The zeolites under static conditions differ in methodology, molar composition, silicon source, and temperature. Adapted from Reference [9,26–29].

The crystallization of MCM-22 is also influenced by the source of silicon used because its degree of dissolution affects nucleation and crystal growth. A study compared three silicon sources with different surface areas: silicic acid (750 m^2·g^{-1}), silica gel (500 m^2·g^{-1}), and precipitated Ultrasil silica (176 m^2·g^{-1}); zeolite with 100% crystallinity was obtained with silicic acid followed by silica gel (90%) and Ultrasil (80%) by aging the gel for 7 days in dynamic conditions [26]. The authors showed high crystallinities obtained under static conditions using silicic acid when compared with the other silicon sources. This indicates that silicon sources with a high surface area are a determining factor in the crystallization of MCM-22. Figure 3a,b show the morphologies of materials synthesized with silicic acid.

Other sources of silicon were used, such as sodium metasilicate, water-glass, and colloidal silica [29]. The use of sodium metasilicate reduced the induction period (less than 12 h) with a crystallized product after 6 days. Colloidal silica and water-glass required induction periods of 2 and 2.5 days, respectively. These differences were attributed to the different dissolution rates of each silicon source. The morphologies of zeolites synthesized with colloidal silica, sodium metasilicate, and water-glass are shown in Figure 3 (images f–h, respectively).

The use of silicon alkoxide as tetraethyl orthosilicate (TEOS) for the synthesis of MCM-22 has been reported [27]. The methodology involves a first step of pre-hydrolysis of TEOS catalyzed with a strong acid media (pH ranging from 0.98–1.65), followed by a second step of hydrothermal reaction of the hydrolyzed precursor with HMI and a source of aluminum in a base media with a pH value ranging from 11–12. This allows a shorter crystallization time, which differs from traditional methods where hydrolysis, condensation, and crystallization occur simultaneously in the same basic medium. According to the authors, MCM-22 with a crystallinity of 98% was produced after 3 days at 158 °C. Figure 3d shows the morphology of the obtained product.

Silica from burned rice husks was used to synthesize MCM-22 [30]. X-ray diffraction analysis confirmed that the product has an MWW structure and textural analysis showed a surface area of 384 m^2·g^{-1} and a pore volume of 0.28 cm^3·g^{-1}. Microscopic analysis showed different particles with interrupted growths, spherical aggregates, and concentric rings.

Structure Directing Agent (SDA)

The design of SDA for the synthesis of zeolites is a subject of continuous research and, for the MWW topology, it is possible to synthesize different materials with other SDAs than HMI. Here, the use of different SDAs is organized in chronological order.

1987—An aluminosilicate-based material was discovered, named SSZ-25, which exhibited the same characteristics as MWW materials [31]. However, it was initially assumed that the material only had 12-ring channels. Subsequently, it was confirmed that the structure of SSZ-25 was isomorphic to the structure of PSH-3 previously synthesized with HMI three years prior [32]. In this case, *N,N,N*-trimethyl-1-adamantyl ammonium hydroxide (TMAda$^+$OH$^-$) was used as an SDA.

1988—A material called ERB-1 was reported, which was the first LZP where aluminum and boron were tetrahedrally coordinated into the MWW structure and piperidine was used as an SDA, and the use of alkali cations was not necessary [33].

1998—TMAda$^+$OH$^-$ was also used to obtain ITQ-1, a pure silica zeolite. To obtain this material, mixtures of TMAda$^+$OH$^-$ with HMI had a particular role in the synthesis because TMAda$^+$OH$^-$ allowed the formation of the external 12-ring hemicavities and HMI contributed to the stabilization of the sinusoidal 10-ring channels present in the internal structure of the MWW lamella [14].

2004—The use of diethyldimethylammonium (DEDMA), ethyltrimethylammonium (ETMA), or hexamethonium (HM) cations as the SDA were reported, and the obtained material was called UZM-8 [34]. UZM-8 was synthesized with a Si/Al molar ratio between 6.5 and 35 and a disordered lamellar structure similar to that of MCM-56 zeolite.

2006—The use of *N*-methylsparteinium (MSPT) as an SDA in a high-throughput synthesis led to the discovery of ITQ-30, an MWW-type zeolite with disordered lamellae similar to MCM-56 [35].

2011—The use of (bis(*N*,*N*,*N*-trimethyl)-1,5-pentanediaminium dibromide as an SDA was reported and conducted to form an EMM-10 precursor [36]. The material is similar to (P)MCM-22, but its lamellae are vertically misaligned.

2013—The use of 1,3-diisopropylimidazolium, a 1,3-diisobutylimidazolium cation, and 1,3-dicyclohexylimidazolium cations were reported to obtain a zeolite named SSZ-70 [37].

2015—A new synthesis of MWW-type materials was reported. It employed 1,3-bis (cyclohexyl)imidazolium hydroxide (IM$^+$OH$^-$) as an SDA, and the obtained materials were called ECNU-5A and ECNU-5B [38]. The procedure used calcined ITQ-1 as a silica source, which was recrystallized with an aqueous solution containing IM$^+$OH$^-$. The crystals of MWW rapidly dissolved due to the high basic pH in only 1 h at 170 °C, yielding only 17.8% and increasing to 92.3% after 24 h. The obtained materials showed a horizontal displacement with misaligned MWW lamellae structure in ABAB or ABC stacking sequence, caused by the geometry between IM$^+$OH$^-$ and the silica structure.

The use of aniline (AN) with mixtures of HMI for the synthesis of MWW zeolites was reported [39,40]. In this case, AN acts as a structure-promoting agent via space filling, and the authors pointed out that the use of AN contributed to the formation of the zeolitic structure because the molecules were not trapped within the MWW structure. Moreover, its recovery and recycling may contribute to low-cost synthesis.

2017—1-adamantanamine as an SDA was reported and the obtained material, called ECNU-10, had a three-dimensional structure analogous to MCM-49 zeolite [41]. ECNU-10 could be obtained when the gel Si/Al ratio was 12–13.5 in a relatively narrow phase region.

2018—A direct synthesis of three-tridimensional MCM-49 using cyclohexylamine (CHA) as an SDA was reported [42]. CHA has a low toxicity and low cost and the obtained results showed that more CHA molecules occupy the hemicavities on the surface and the supercavities, and the SiO$_2$/Al$_2$O$_3$ ratio of the obtained product could be up to 34.6. The authors compared the products synthesized with CHA or HMI for the liquid phase alkylation of benzene with ethylene and similar catalytic performances were observed.

3. MWW-Type Materials by Post-Synthesis Modifications

(P)MCM-22 offers diverse possibilities to obtain more open structures. The first example of this is the interlayer expanded zeolite called IEZ-MWW, in which silanol groups of the LZP were reacted with alkoxysilanes such as SiMe$_2$Cl$_2$ or Si(EtO)$_2$Me$_2$ [43]. After calcination, a 12-ring pore was formed by the single silicon atoms that act as small pillars, as represented in Figure 4. The increase in pore structure may serve catalytic purposes to diffuse large molecules and as selective adsorbents for adsorption and separation.

The successful swelling procedure is a key step to obtain a hybrid organic-inorganic material used to form pillared and delaminated materials with high accessibility. In contrast, this procedure is still challenging (cost- and time-consuming and with the possible formation of competing mesophases). The separation of individual MWW lamellae was carried out using long alkyl organic molecules (hexadecyltrimethylammonium cations, CTA$^+$, usually) to populate the interlamellar region. An alkaline media was needed to deprotonate the silanol groups and break the hydrogen bonds between the lamellae. Tetrapropylammonium hydroxide is a double agent because it supplies hydroxide ions and its counter ion (TPA$^+$) to facilitate the entering of the CTA$^+$ molecules into the interlamellar region. When NaOH is used, the small Na$^+$ cations rapidly enter the interlamellar region and compensate negatively charged ions before populating the CTA$^+$ molecules in the interlamellar region, resulting in an unsuccessfully or partially swollen material [44]. Several studies have sought to better understand swollen materials using different swelling conditions (room temperature or 80 °C), molecular dimensions of swelling agents, hydroxide sources, and strategies of recycling and reusing the swelling solution [45–50].

MCM-36 was the first pillared molecular sieve with zeolite properties. The swollen precursor was mixed with a pillaring agent (TEOS) and went through subsequent calcination where rigid silica pillars formed, keeping the individual MWW lamella separated from each other. Characterization results showed a surface area of 896 m$^2 \cdot$g^{-1} (compared with 400 m$^2 \cdot$g^{-1} for MCM-22) with mesopores between 2 and 4 nm and higher adsorption capacities of bulky molecules as 1,3,5-trimethylbenze (TMB) with 0.040 mg$\cdot$g^{-1}, whereas MCM-22 showed negligible adsorption [51,52].

Another important pillared material was reported, but in this case, aryl silsesquioxane molecules acted as organic pillars between MWW lamellae to obtain a multifunctional organic-inorganic catalytic material with a hierarchical structure [53]. The swollen precursor was reacted with a solution of 1,4-bis(triethoxysilyl)benzene (BTEB) and the CTA$^+$ molecules were removed by acid extraction. Following this, amino groups were incorporated onto the bridged benzene groups in the interlayer space. The obtained materials showed acid sites provided by the MWW structure of lamellae combined with basic sites from amino groups incorporated on the aryl molecules. Characterization results showed a basal spacing of 4.1 nm, a surface area of 556 m$^2 \cdot$g^{-1}, and a mesoporous region formed by the separated lamellae.

The use of swollen MWW materials treated with ultrasound and acidic medium and a posterior calcination generate ITQ-2, the first delaminated zeolite [54]. Its surface area showed 700 m$^2 \cdot$g^{-1} and a broad distribution of mesopores due to the random stacking of MWW lamellae in edge-to-face orientation, as shown in Figure 4. ITQ-2 had superior capacities (7 times higher than MCM-22) for the adsorption of TMB and superior catalytic performance for reactions with the cracking of bulky molecules, such as 1,3-diisopropylbenzene and vacuum gas oil in gasoline and diesel [55].

The use of confined subnanometric platinum species was reported and made use of a swelling procedure [56]. In this approach, a solution containing subnanometric platinum species was added during the swelling procedure of the ITQ-1 precursor. After calcination, a three-dimensional zeolite containing platinum confined in the supercavities and on the external surface of the MWW crystallites was formed, as shown in Figure 5. The authors also studied the growth of these platinum species by high-temperature oxidation-reduction treatments, obtaining small nanoparticles with sizes between 1 and 2 nm.

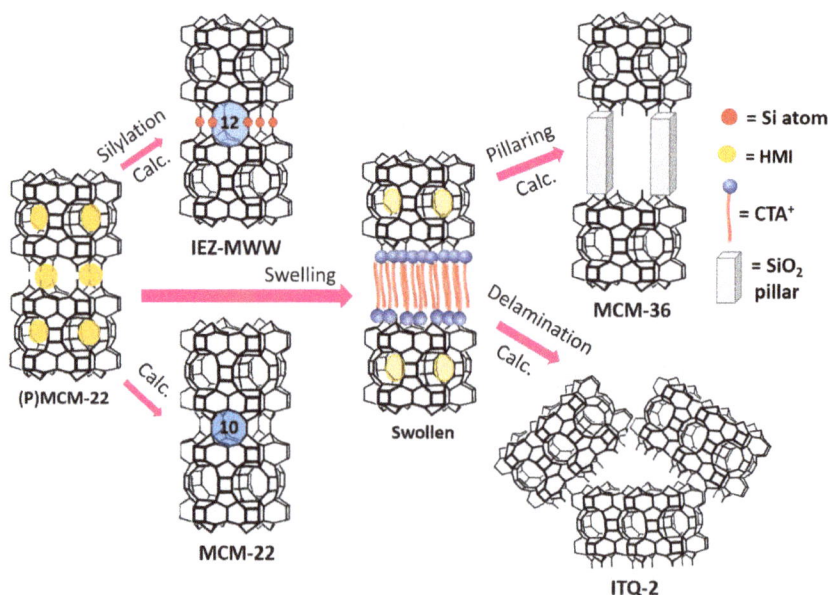

Figure 4. Representation of the postsynthetic procedures to obtain the interlayer expanded zeolite IEZ-MWW, swollen, pillared (MCM-36), and delaminated (ITQ-2) MWW-type materials.

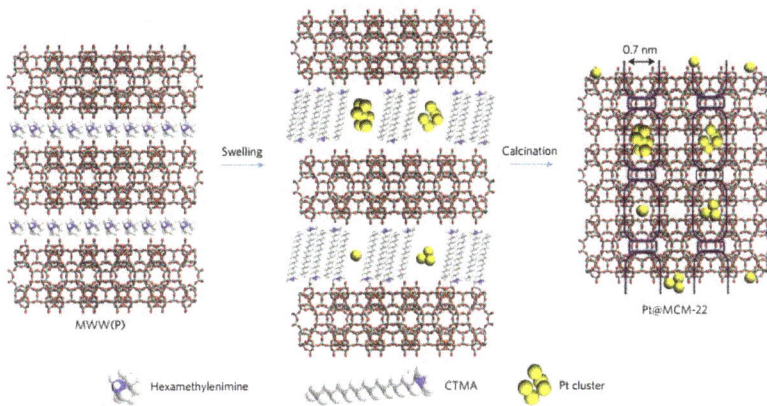

Figure 5. Scheme of confining platinum species in the MCM-22 structure by swelling the MWW precursor with surfactants and platinum species and a subsequent calcination. Reprinted with permission from Springer Nature, Reference [56].

Recently, a novel strategy to obtain delaminated MWW-type zeolite employed a treatment using commercially available telechelic liquid polybutadienes at room temperature and a swollen precursor [57]. The resulting swollen precursor/polymer suspension was subject to ultrasound or a chaotic flow treatment in a planetary mixing system, as shown in Figure 6. The authors confirmed delamination using small angle X-ray scattering (SAXS) results, where no interlamellar reflections were observed using hydroxyl-terminated polybutadiene (HTPB) with 36 min of chaotic flow. On the other hand, the sonication procedure is also effective, but required 5 h to obtain a delaminated material. Another interesting result was the increase in the interlamellar space of the swollen precursor from 4.6 nm to 9.4 nm after manual mixing with HTPB for only 1 min. The authors also pointed out that the end groups of the liquid polybutadienes preferentially interact with the zeolite surface through hydrogen bonds and this is the key factor needed to obtain a delaminated material.

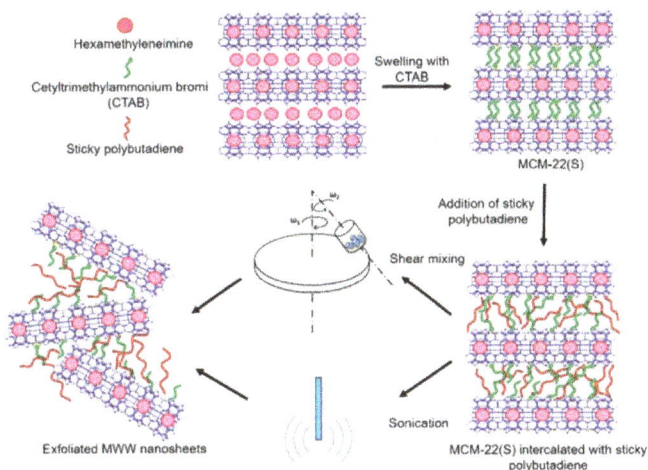

Figure 6. Scheme of the formation of delaminated MWW-type material using telechelic liquid polybutadienes. Adapted from Reference [57]. Copyright (2017), with permission of The Royal Society of Chemistry.

Another post-synthetic approach to obtain more open structures in MWW-type materials is to generate intracrystalline mesopores [41]. This strategy uses MCM-49 zeolite and mixtures of CTA^+ and NaOH under different temperatures and times to obtain desilicated MWW-type materials, as shown in Figure 7. Under the post-synthesis treatment, fragments of the MWW zeolitic structure were removed by the attack of the hydroxide ions in the defects, whereas the CTA^+ molecules acted as a defensive barrier to avoid uncontrollable dissolution by NaOH. Thus, intracrystalline mesopores were formed by the regions where CTA^+ presented "defensive failures." The obtained materials showed a distribution of mesopores with sizes between 2 and 4 nm.

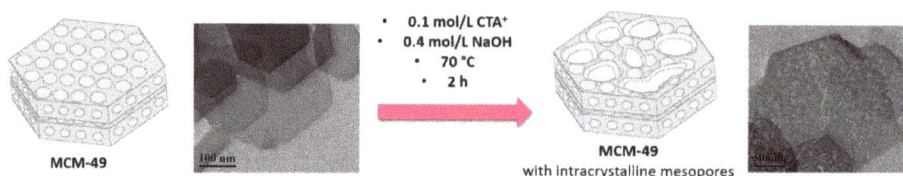

Figure 7. Post-synthesis treatment to obtain intracrystalline mesopores in MCM-49 zeolite. TEM micrographs were taken from Reference [58]. Copyright (2015), with permission from Elsevier.

4. MWW-Type Materials by Direct Synthesis

Several efforts have been made to obtain individual zeolitic lamellae separated by direct synthesis, with the main aim of eliminating the swelling and delamination steps. The first MWW-type material with individual lamellae separated from each other was called Direct Synthesis ITQ-2 (DS-ITQ-2). DS-ITQ-2 was obtained by a very similar traditional synthesis of (P)MCM-22 that was modified using N-hexadecyl-N'-methyl-DABCO ($C_{16}DC_1$) as a dual template, as shown in Figure 8. DABCO acted as an SDA because of its similarity to HMI, while the tail group (C_{16}) avoided the stacking and growth of the structure along the c-axis. The calcined material showed a surface area of 545 $m^2 \cdot g^{-1}$ and microporous volume of 0.12 $cm^3 \cdot g^{-1}$, which is higher than that of the traditional ITQ-2 (0.08 $cm^3 \cdot g^{-1}$). The pore size distribution showed values in a broad range, which is characteristic of delaminated-type zeolites.

MIT-1 was another MWW-type delaminated material obtained by direct synthesis [59]. However, the organic molecule was the $TMAda^+OH^-$ linked by alkyl chains with four, five, or six carbons and connected with a CTA^+ molecule, as shown in Figure 8. In this case, the adamantylammonium as the head-group acted as an SDA, the linkers acted to stabilize the pore mouth, and the hydrophobic tail of the CTA^+ molecule prevented zeolitic growth along the c-axis. MIT-1 had a surface area higher than 500 $m^2 \cdot g^{-1}$ and a broad distribution of mesopore sizes.

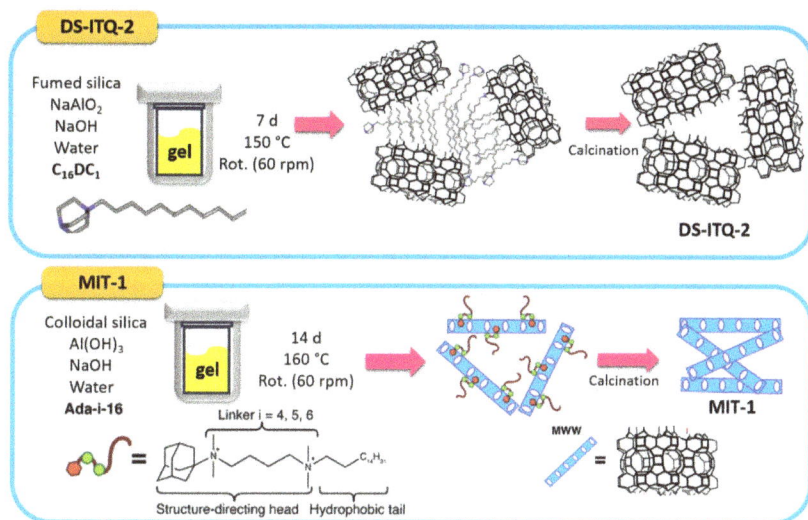

Figure 8. Representation of the syntheses of DS-ITQ-2 and MIT-1.

Both DS-ITQ-2 and MIT-1 are examples of MWW-type materials obtained when the SDA was linked with a long hydrophobic alkyl chain. Another strategy made use of mixtures of SDA and CTA$^+$ using a dissolution-recrystallization route to obtain a swollen precursor called Al-ECNU-7P by direct crystallization [60]. The synthesis comprises a dissolution containing MWW seeds (ITQ-1), the 1,3-bis(cyclohexyl)imidazolium hydroxide as an SDA, and a silicate source at 140 °C for 1 h followed by the addition of CTA$^+$ and the aluminum source. Then, the obtained gel was crystallized at 150 °C for 72 h, as shown in Figure 9. From the Al-ECNU-7P, it was possible to obtain a pillared material with a basal spacing of 5 nm, a surface area of 701 m$^2 \cdot$g^{-1}, and a mesopore size distribution centered at 3.1 nm. The direct calcination of Al-ECNU-7P showed delaminated and partially condensed lamellae, a surface area of 502 m$^2 \cdot$g^{-1}, and a mesopore size distribution centered at 5 nm.

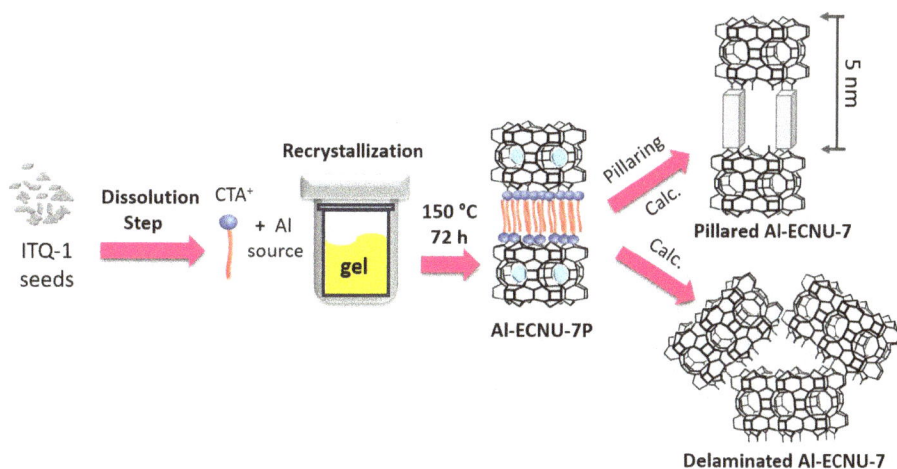

Figure 9. General scheme of the synthesis of Al-ECNU-7P as well as its pillared and delaminated forms.

Another strategy to overcome the diffusional barrier imposed on reactants and products is the synthesis of nanosized zeolites. The decrease of common microsized zeolite crystals to nanosized crystals increases the external surface and could facilitate the rapid diffusion of reactants and products [61]. The synthesis of nanosized MCM-22 zeolite was reported using polydiallydimethylammonium chloride (PDDA) as a protecting or stabilizing agent to avoid the self-aggregation and intergrowth of silica colloids by direct synthesis, as shown in the scheme of Figure 10a. The self-assembly of the cationic polymer and the negatively charged silica species interactions are the main reasons to obtain nanosized MCM-22 with a crystal size of 40 nm, as shown in Figure 10b.

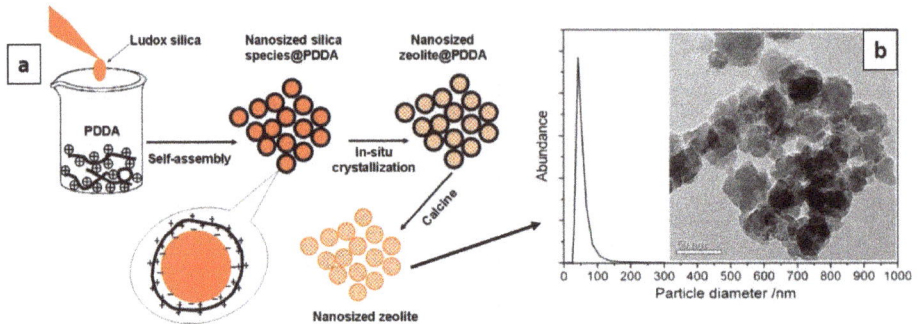

Figure 10. Scheme of the synthesis of nanosized MCM-22 using PDDA (**a**); particle size distribution and TEM of the obtained crystals (**b**). Adapted from Reference [62]. Copyright (2014), with permission from Elsevier.

Recently, a direct synthesis method was reported to obtain dandelion-like MCM-22 microspheres with interparticle meso/macro voids [63]. The authors used carbon black pearls (BP 2000) as a hard template and the synthesis is shown in Figure 11. In the synthesis procedure, colloidal silica was slowly added dropwise to a mixture containing water, sodium aluminate, HMI, and BP 2000 in rotation during nucleation and crystal growth.

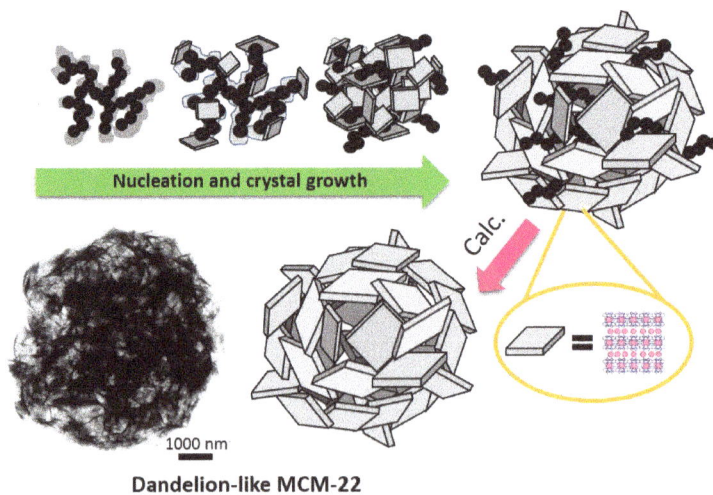

Figure 11. Proposed mechanism of the formation of dandelion-like MCM-22 microspheres. Adapted from Reference [63].

The interactions between BP 2000 and the gel precursor were due to the hydrogen bonds between the functional groups (carboxylic acid, ketone, and ester) present in carbon black and the silanol and amino groups derived from HMI, as well as the $Si-O$ and $Al-O$ bonds. The tortuous shape of BP 2000 aggregates interacts only with the external surfaces of the MWW crystals, forming thin MWW crystal platelets stacked in edge-to-face orientations with interparticle porosity. The pore size distribution showed large mesopores and macropores centered at 200 nm, which is two times higher than that of the traditional MCM-22 zeolite.

5. Conclusions

MWW-type zeolites are attractive nanoporous materials for different applications, due to their three-dimensional and two-dimensional forms. It is interesting that, even 30 years after the discovery of materials with MWW topology, research and development around this family is a matter of continuous interest. Most of this is due to the versatility of the lamellar zeolitic precursor that allows the creation of elegant and different pore architectures with tunable physicochemical properties in terms of acidity, accessibility, and structural stability. The direct synthesis of MWW-type materials with different lamellae organizations are directly linked to the synthesis and discovery of new SDAs with a special attention paid to the dual templates that avoid excessive growth and stacking of the structure along the c-axis. In the future, these routes of synthesis may gain more prominence and extend to other zeolitic structures.

Author Contributions: Conceptualization, A.S; Writing-Original Draft Preparation, A.S and S.P.; Writing-Review & Editing, A.S; Supervision, S.P.

Funding: This research received no external funding.

Acknowledgments: Anderson Schwanke thanks the CAPES Foundation and INOMAT (project number: 88887.136344/2017-00 - 465452/2014-0).

Conflicts of Interest: The authors declare no conflict of interest.

References

1. Tanabe, K.; Hölderich, W.F. Industrial application of solid acid–base catalysts. *Appl. Catal. A Gen.* **1999**, *181*, 399–434. [CrossRef]
2. Vartuli, J.C.; Degnan, T.F., Jr. Applications of mesoporous molecular sieves in catalysis and separations. In *Studies in Surface Science and Catalysis*; Jiří Čejka, H.v.B.A.C., Ferdi, S., Eds.; Elsevier: New York, NY, USA, 2007; Volume 168, pp. 837–854.
3. Corma, A. From Microporous to Mesoporous Molecular Sieve Materials and Their Use in Catalysis. *Chem. Rev.* **1997**, *97*, 2373–2420. [CrossRef] [PubMed]
4. Rinaldi, R.; Schuth, F. Design of solid catalysts for the conversion of biomass. *Energy Environ. Sci.* **2009**, *2*, 610–626. [CrossRef]
5. International Zeolite Association (IZA). Available online: http://www.iza-structure.org/ (accessed on 20 June 2018).
6. Roth, W.J.; Gil, B.; Makowski, W.; Marszalek, B.; Eliasova, P. Layer like porous materials with hierarchical structure. *Chem. Soc. Rev.* **2016**, *45*, 3400–3438. [CrossRef] [PubMed]
7. Diaz, U.; Corma, A. Layered zeolitic materials: An approach to designing versatile functional solids. *Dalton Trans.* **2014**, *43*, 10292–10316. [CrossRef] [PubMed]
8. Roth, W.J.; Cejka, J. Two-dimensional zeolites: Dream or reality? *Catal. Sci. Technol.* **2011**, *1*, 43–53. [CrossRef]
9. Ramos, F.S.O.; de Pietre, M.K.; Pastore, H.O. Lamellar zeolites: An oxymoron? *RSC Adv.* **2013**, *3*, 2084–2111. [CrossRef]
10. Opanasenko, M.V.; Roth, W.J.; Cejka, J. Two-dimensional zeolites in catalysis: Current status and perspectives. *Catal. Sci. Technol.* **2016**, *6*, 2467–2484. [CrossRef]
11. Schwanke, A.J.; Pergher, S. Hierarchical MWW Zeolites by Soft and Hard Template Routes. In *Handbook of Ecomaterials*; Martínez, L.M.T., Kharissova, O.V., Kharisov, B.I., Eds.; Springer: Berlin, Germany, 2017; pp. 1–23.

12. Rubin, M.K.; Chu, P. Composition of Synthetic Porous Crystalline Material, Its Synthesis and Use. U.S. Patent 4954325A, 4 September 1990.

13. Leonowicz, M.E.; Lawton, J.A.; Lawton, S.L.; Rubin, M.K. MCM-22: A Molecular Sieve with Two Independent Multidimensional Channel Systems. *Science* **1994**, *264*, 1910–1913. [CrossRef] [PubMed]

14. Camblor, M.A.; Corma, A.; Díaz-Cabañas, M.-J.; Baerlocher, C. Synthesis and Structural Characterization of MWW Type Zeolite ITQ-1, the Pure Silica Analog of MCM-22 and SSZ-25. *J. Phys. Chem. B* **1998**, *102*, 44–51. [CrossRef]

15. Zhou, D.; Bao, Y.; Yang, M.; He, N.; Yang, G. DFT studies on the location and acid strength of Brönsted acid sites in MCM-22 zeolite. *J. Mol. Catal. A Chem.* **2006**, *244*, 11–19. [CrossRef]

16. Lawton, S.L.; Fung, A.S.; Kennedy, G.J.; Alemany, L.B.; Chang, C.D.; Hatzikos, G.H.; Lissy, D.N.; Rubin, M.K.; Timken, H.-K.C.; Steuernagel, S.; et al. Zeolite MCM-49: A Three-Dimensional MCM-22 Analogue Synthesized by in Situ Crystallization. *J. Phys. Chem. B* **1996**, *100*, 3788–3798. [CrossRef]

17. Bennett, J.M.; Lawton, C.D.C.S.L.; Leonowicz, M.E.; Lissy, D.N.; Rubin, M.K. Synthetic porous crystalline MCM-49, its synthesis and use. U.S. Patent 5236575, 17 August 1993.

18. Fung, A.S.; Lawton, S.L.; Roth, W.J. Synthetic Layered MCM-56, Its Synthesis and Use. U.S. Patent 5362697A, 8 November 1994.

19. Delitala, C.; Alba, M.D.; Becerro, A.I.; Delpiano, D.; Meloni, D.; Musu, E.; Ferino, I. Synthesis of MCM-22 zeolites of different Si/Al ratio and their structural, morphological and textural characterisation. *Microporous Mesoporous Mater.* **2009**, *118*, 1–10. [CrossRef]

20. Ravishankar, R.; Sen, T.; Ramaswamy, V.; Soni, H.S.; Ganapathy, S.; Sivasanker, S. Synthesis, Characterization and Catalytic properties of Zeolite PSH-3/MCM-22. In *Studies in Surface Science and Catalysis*; Weitkamp, J., Karge, H.G., Pfeifer, H., Hölderich, W., Eds.; Elsevier: New York, NY, USA, 1994; Volume 84, pp. 331–338.

21. Corma, A.; Corell, C.; Pérez-Pariente, J.; Guil, J.M.; Guil-López, R.; Nicolopoulos, S.; Calbet, J.G.; Vallet-Regi, M. Adsorption and catalytic properties of MCM-22: The influence of zeolite structure. *Zeolites* **1996**, *16*, 7–14. [CrossRef]

22. Wang, Y.M.; Shu, X.T.; He, M.Y. 02-P-34—Static synthesis of zeolite MCM-22. In *Studies in Surface Science and Catalysis*; Galarneau, A., Fajula, F., Di Renzo, F., Vedrine, J., Eds.; Elsevier: New York, NY, USA, 2001; Volume 135, p. 194.

23. Bevilacqua, M.; Meloni, D.; Sini, F.; Monaci, R.; Montanari, T.; Busca, G. A Study of the Nature, Strength, and Accessibility of Acid Sites of H-MCM-22 Zeolite. *J. Phys. Chem. C* **2008**, *112*, 9023–9033. [CrossRef]

24. Meloni, D.; Laforge, S.; Martin, D.; Guisnet, M.; Rombi, E.; Solinas, V. Acidic and catalytic properties of H-MCM-22 zeolites: 1. Characterization of the acidity by pyridine adsorption. *Appl. Catal. A* **2001**, *215*, 55–66. [CrossRef]

25. Onida, B.; Geobaldo, F.; Testa, F.; Aiello, R.; Garrone, E. H-Bond Formation and Proton Transfer in H-MCM-22 Zeolite as Compared to H-ZSM-5 and H-MOR: An FTIR Study. *J. Phys. Chem. B* **2002**, *106*, 1684–1690. [CrossRef]

26. Güray, I.; Warzywoda, J.; Baç, N.; Sacco, A., Jr. Synthesis of zeolite MCM-22 under rotating and static conditions. *Microporous Mesoporous Mater.* **1999**, *31*, 241–251. [CrossRef]

27. Wu, Y.; Ren, X.; Wang, J. Facile synthesis and morphology control of zeolite MCM-22 via a two-step sol–gel route with tetraethyl orthosilicate as silica source. *Mater. Chem. Phys.* **2009**, *113*, 773–779. [CrossRef]

28. Inagaki, S.; Kamino, K.; Hoshino, M.; Kikuchi, E.; Matsukata, M. Textural and catalytic properties of MCM-22 zeolite crystallized by the vapor-phase transport method. *Bull. Chem. Soc. Jpn.* **2004**, *77*, 1249–1254. [CrossRef]

29. Wu, Y.; Ren, X.; Lu, Y.; Wang, J. Crystallization and morphology of zeolite MCM-22 influenced by various conditions in the static hydrothermal synthesis. *Microporous Mesoporous Mater.* **2008**, *112*, 138–146. [CrossRef]

30. Cheng, Y.; Lu, M.; Li, J.; Su, X.; Pan, S.; Jiao, C.; Feng, M. Synthesis of MCM-22 zeolite using rice husk as a silica source under varying-temperature conditions. *J. Colloid Interface Sci.* **2012**, *369*, 388–394. [CrossRef] [PubMed]

31. Zones, S.I. A crystalline zeolite SSZ-25 Is Prepared Using an Adamantane Quaternary Ammonium Ion as a Template. E.U. Patent 231860, 12 May 1987.

32. Puppe, L.; Weisser, J. Crystalline Aluminosilicate PSH-3 and Its Process of Preparation. U.S. Patent 4439409A, 23 March 1984.

33. Bellussi, G.; Perego, G.; Cierici, M.G.; Giusti, A. Bulletin of the Chemical Society of Japan. E.U. Patent 293032, 30 November 1988.

34. Rohde, L.M.; Lewis, G.J.; Miller, M.A.; Moscoso, J.G.; Gisselquist, J.L.; Patton, R.L.; Wilson, S.T.; Jan, D.Y. Crystalline Aluminosilicate Zeolitic Composition: UZM-8. U.S. Patent 6756030B1, 21 March 2003.

35. Corma, A.; Díaz-Cabanas, M.J.; Moliner, M.; Martínez, C. Discovery of a new catalytically active and selective zeolite (ITQ-30) by high-throughput synthesis techniques. *J. Catal.* **2006**, *241*, 312–318. [CrossRef]

36. Roth, W.J.; Dorset, D.L.; Kennedy, G.J. Discovery of new MWW family zeolite EMM-10: Identification of EMM-10P as the missing MWW precursor with disordered layers. *Microporous Mesoporous Mater.* **2011**, *142*, 168–177. [CrossRef]

37. Zones, S.I.; Davis, T.M. Zeolite SSZ-70 Having Enhanced External Surface Area. E.U. Patent 3027559B1, 30 July 2013.

38. Xu, L.; Ji, X.; Jiang, J.-G.; Han, L.; Che, S.; Wu, P. Intergrown Zeolite MWW Polymorphs Prepared by the Rapid Dissolution–Recrystallization Route. *Chem. Mater.* **2015**, *27*, 7852–7860. [CrossRef]

39. Xing, E.; Shi, Y.; Xie, W.; Zhang, F.; Mu, X.; Shu, X. Synthesis, characterization and application of MCM-22 zeolites via a conventional HMI route and temperature-controlled phase transfer hydrothermal synthesis. *RSC Adv.* **2015**, *5*, 8514–8522. [CrossRef]

40. Xing, E.; Shi, Y.; Xie, W.; Zhang, F.; Mu, X.; Shu, X. Perspectives on the multi-functions of aniline: Cases from the temperature-controlled phase transfer hydrothermal synthesis of MWW zeolites. *Microporous Mesoporous Mater.* **2017**, *254*, 201–210. [CrossRef]

41. Ji, P.; Shen, M.; Lu, K.; Hu, B.; Jiang, J.-G.; Xu, H.; Wu, P. ECNU-10 zeolite: A three-dimensional MWW-Type analogue. *Microporous Mesoporous Mater.* **2017**, *253*, 137–145. [CrossRef]

42. Chu, W.; Li, X.; Liu, S.; Zhu, X.; Xie, S.; Chen, F.; Wang, Y.; Xin, W.; Xu, L. Direct synthesis of three-dimensional MWW zeolite with cyclohexylamine as an organic structure-directing agent. *J. Mat. Chem. A* **2018**. [CrossRef]

43. Wu, P.; Ruan, J.; Wang, L.; Wu, L.; Wang, Y.; Liu, Y.; Fan, W.; He, M.; Terasaki, O.; Tatsumi, T. Methodology for Synthesizing Crystalline Metallosilicates with Expanded Pore Windows Through Molecular Alkoxysilylation of Zeolitic Lamellar Precursors. *J. Am. Chem. Soc.* **2008**, *130*, 8178–8187. [CrossRef] [PubMed]

44. Roth, W.J. Cation Size Effects in Swelling of the Layered Zeolite Precursor MCM-22-P. *Pol. J. Chem.* **2006**, *80*, 703–708.

45. Maheshwari, S.; Jordan, E.; Kumar, S.; Bates, F.S.; Penn, R.L.; Shantz, D.F.; Tsapatsis, M. Layer Structure Preservation during Swelling, Pillaring, and Exfoliation of a Zeolite Precursor. *J. Am. Chem. Soc.* **2008**, *130*, 1507–1516. [CrossRef] [PubMed]

46. Schwanke, A.J.; Pergher, S.; Díaz, U.; Corma, A. The influence of swelling agents molecular dimensions on lamellar morphology of MWW-type zeolites active for fructose conversion. *Microporous Mesoporous Mater.* **2017**, *254*, 17–27. [CrossRef]

47. Chlubná, P.; Roth, W.J.; Zukal, A.; Kubů, M.; Pavlatová, J. Pillared MWW zeolites MCM-36 prepared by swelling MCM-22P in concentrated surfactant solutions. *Catal. Today* **2012**, *179*, 35–42. [CrossRef]

48. Roth, W.J.; Chlubná, P.; Kubů, M.; Vitvarová, D. Swelling of MCM-56 and MCM-22P with a new medium—Surfactant-tetramethylammonium hydroxide mixtures. *Catal. Today* **2013**, *204*, 8–14. [CrossRef]

49. Roth, W.J.; Čejka, J.; Millini, R.; Montanari, E.; Gil, B.; Kubu, M. Swelling and Interlayer Chemistry of Layered MWW Zeolites MCM-22 and MCM-56 with High Al Content. *Chem. Mater.* **2015**, *27*, 4620–4629. [CrossRef]

50. Schwanke, A.J.; Díaz, U.; Corma, A.; Pergher, S. Recyclable swelling solutions for friendly preparation of pillared MWW-type zeolites. *Microporous Mesoporous Mater.* **2017**, *253*, 91–95. [CrossRef]

51. Kresge, C.T.; Roth, W.J.; Simmons, K.G.; Vartuli, J.C. Layered oxide materials and swollen and pillared forms thereof. WO Patent 1992011934A1, 23 July 1992.

52. Roth, W.J.; Kresge, C.T.; Vartuli, J.C.; Leonowicz, M.E.; Fung, A.S.; McCullen, S.B. MCM-36: The first pillared molecular sieve with zeolite properties. In *Studies in Surface Science and Catalysis*; Beyer, H.K., Karge, H.G., Kiricsi, I., Nagy, J.B., Eds.; Elsevier: New York, NY, USA, 1995; Volume 94, pp. 301–308.

53. Corma, A.; Díaz, U.; García, T.; Sastre, G.; Velty, A. Multifunctional Hybrid Organic−Inorganic Catalytic Materials with a Hierarchical System of Well-Defined Micro- and Mesopores. *J. Am. Chem. Soc.* **2010**, *132*, 15011–15021. [CrossRef] [PubMed]

54. Corma, A.; Fornes, V.; Pergher, S.B.C.; Maesen, T.L.M.; Buglass, J.G. Delaminated zeolite precursors as selective acidic catalysts. *Nature* **1998**, *396*, 353–356. [CrossRef]

55. Corma, A.; Diaz, U.; Fornés, V.; Guil, J.M.; Martínez-Triguero, J.; Creyghton, E.J. Characterization and Catalytic Activity of MCM-22 and MCM-56 Compared with ITQ-2. *J. Catal.* **2000**, *191*, 218–224. [CrossRef]

56. Liu, L.; Díaz, U.; Arenal, R.; Agostini, G.; Concepción, P.; Corma, A. Corrigendum: Generation of subnanometric platinum with high stability during transformation of a 2D zeolite into 3D. *Nat. Mat.* **2017**, *16*, 1272. [CrossRef] [PubMed]

57. Sabnis, S.; Tanna, V.A.; Li, C.; Zhu, J.; Vattipalli, V.; Nonnenmann, S.S.; Sheng, G.; Lai, Z.; Winter, H.H.; Fan, W. Exfoliation of two-dimensional zeolites in liquid polybutadienes. *Chem. Commun.* **2017**, *53*, 7011–7014. [CrossRef] [PubMed]

58. Gao, N.; Xie, S.; Liu, S.; Xin, W.; Gao, Y.; Li, X.; Wei, H.; Liu, H.; Xu, L. Development of hierarchical MCM-49 zeolite with intracrystalline mesopores and improved catalytic performance in liquid alkylation of benzene with ethylene. *Microporous Mesoporous Mater.* **2015**, *212*, 1–7. [CrossRef]

59. Luo, H.Y.; Michaelis, V.K.; Hodges, S.; Griffin, R.G.; Roman-Leshkov, Y. One-pot synthesis of MWW zeolite nanosheets using a rationally designed organic structure-directing agent. *Chem. Sci.* **2015**, *6*, 6320–6324. [CrossRef] [PubMed]

60. Xu, L.; Ji, X.; Li, S.; Zhou, Z.; Du, X.; Sun, J.; Deng, F.; Che, S.; Wu, P. Self-Assembly of Cetyltrimethylammonium Bromide and Lamellar Zeolite Precursor for the Preparation of Hierarchical MWW Zeolite. *Chem. Mater.* **2016**, *28*, 4512–4521. [CrossRef]

61. Mintova, S.; Grand, J.; Valtchev, V. Nanosized zeolites: Quo Vadis? *C. R. Chim.* **2016**, *19*, 183–191. [CrossRef]

62. Yin, X.; Chu, N.; Yang, J.; Wang, J.; Li, Z. Synthesis of the nanosized MCM-22 zeolite and its catalytic performance in methane dehydro-aromatization reaction. *Catal. Commun.* **2014**, *43*, 218–222. [CrossRef]

63. Schwanke, A.; Villarroel-Rocha, J.; Sapag, K.; Díaz, U.; Corma, A.; Pergher, S. Dandelion-Like Microspherical MCM-22 Zeolite Using BP 2000 as a Hard Template. *ACS Omega* **2018**, *3*, 6217–6223. [CrossRef] [PubMed]

© 2018 by the authors. Licensee MDPI, Basel, Switzerland. This article is an open access article distributed under the terms and conditions of the Creative Commons Attribution (CC BY) license (http://creativecommons.org/licenses/by/4.0/).

![applied sciences logo] *applied sciences*

MDPI

Article

Effects of Different Variables on the Formation of Mesopores in Y Zeolite by the Action of CTA$^+$ Surfactant

Juliana F. Silva, Edilene Deise Ferracine and Dilson Cardoso *

Catalysis Laboratory LabCat, Department of Chemical Engineering, Federal University of São Carlos, P.O. Box 676, São Carlos SP CEP 13.565-905, Brazil; julianafloriano75@gmail.com (J.F.S.); edilenedeise@ufscar.br (E.D.F.)
* Correspondence: dilson@ufscar.br; Tel.: +55-016-3351-8693

Received: 25 June 2018; Accepted: 26 July 2018; Published: 4 August 2018

Abstract: Zeolites are microporous crystalline aluminosilicates with a number of useful properties including acidity, hydrothermal stability, and structural selectivity. However, the exclusive presence of micropores restricts diffusive mass transport and reduces the access of large molecules to active sites. In order to resolve this problem, mesopores can be created in the zeolite, combining the advantages of microporous and mesoporous materials. In this work, mesospores were created in the Ultrastable USY zeolite (silicon/aluminum ratio of 15) using alkaline treatment (NaOH) in the presence of cetyltrimethylammonium bromide surfactant, followed by hydrothermal treatment. The effects of the different concentrations of NaOH and the surfactant on the textural, chemical, and morphological characteristics of the modified zeolites were evaluated. Generating mesoporosity in the USY zeolite was possible through the simultaneous presence of surfactant and alkaline solution. Among the parameters studied, the concentration of the alkaline medium had the greatest influence on the textural properties of the zeolites. The presence of Cetyltrimethylammonium Bromide (CTA$^+$) prevented the amorphization of the structure during the modification and also avoided desilication of the zeolite.

Keywords: zeolites; mesopores; diffusion; surfactant

---

1. Introduction

Zeolites are a class of natural and synthetic minerals that have several structural characteristics in common based on three-dimensional combinations of tetrahedra (TO$_4$, where T represents atoms of silicon and aluminum) connected by oxygen atoms [1].

The microporous zeolite structure is responsible for the acidity, hydrothermal stability, and structural selectivity of these materials [2]. However, the exclusive presence of micropores limits diffusive mass transport and reduces the access of large molecules to the active sites [3,4]. One method for resolving this problem is creating mesopores in the zeolite structure to combine the properties of microporous and mesoporous zeolites in a single material.

Ultrastable USY is a synthetic zeolite is widely used in the petroleum industry. This zeolite is not directly obtained by hydrothermal synthesis, but instead by vapor treatment of Y zeolite. This process creates mesopores that do not significantly affect intra-crystalline diffusion, since they mainly exist as cavities that are connected to the surface by micropores [5,6]. This means that the preparation of USY zeolite possessing secondary porosity has to be performed via post-synthesis modifications.

Compared to conventional zeolites, zeolites with mesopores provide high molecular weight reagents greater access to the active sites. They have shorter retention times of reaction products in the micropores, avoiding secondary reactions and improving selectivity toward the primary products of interest. Furthermore, relative to purely mesoporous materials, they exhibit better hydrothermal stability and higher acidity [7].

The strategies that have been developed to create mesoporosity in zeolites can be grouped into constructive and destructive techniques. In the constructive approach (bottom-up), the mesopores are created during the synthesis of the zeolite, with or without the use of templates. In the destructive route (top-down), mesopores are produced by means of post-synthesis treatments, such as dealumination and desilication. The modification of zeolites using surfactants together with alkaline treatment for the formation of mesoporosity is a highly attractive post-synthesis route, since it produces materials with controlled mesoporosity in terms of the shape, size, connectivity, and location of the mesopores [5,8].

Different approaches have been described for the creation of mesoporosity using surfactants. The recrystallization technique reported by Ivanova et al. [9] involves two stages: the zeolite is first partially destroyed using an alkaline treatment, then the surfactant is added to the reaction mixture, and hydrothermal treatment is applied. This work investigated the creation of mesopores in mordenite (MOR) zeolite using the recrystallization method. Modification of MOR (Si/Al = 49) was performed by treatment with NaOH at room temperature, followed by hydrothermal treatment at 100 °C, in the presence of cetyltrimethylammonium bromide (CTAB). The materials obtained under the conditions used, which were classified as micro- and mesoporous composites, were tested in a transalkylation reaction of biphenyl (BP) with para-diisopropy lbenzene (p-DIPB) and in the cracking reaction of 1,3,5-triisopropylbenzene (TIPB). Each of these reactions required a certain volume of mesopores to ensure enhanced performance.

Another approach reported in the literature is alkaline treatment in the presence of a CTAB surfactant at room temperature, followed by heating of the mixture at 150 °C for several hours to rearrange the structure and create ordered mesopores in the zeolite [8,10,11]. The advantage of this technique, compared to the recrystallization method, is that it avoids complete amorphization of the zeolite by the surfactant, which can occur during more severe alkaline treatments [5].

The crystal rearrangement associated with the use of a surfactant during alkaline treatment was first reported in a patent published in 2005 by Garcia-Martínez [12]. This technique, described as surfactant-templating, was introduced to reduce the difficulty of preparing zeolites with mesopores, instead of materials composed of distinct regions (one mesoporous and the other zeolitic).

USY zeolites with low contents of aluminum are sensitive to treatment with alkaline solution and readily undergo amorphization. For this reason, organic cations such as TPA^+, TMA^+, and CTA^+ alkylammonium cations can be used in the alkaline solution in order to preserve the micropore volume and the associated properties of the zeolite [13,14].

In this work, USY zeolite (Si/Al = 15) was modified by alkaline treatment in the presence of CTAB surfactant, with subsequent hydrothermal treatment to form a substantial volume of mesopores. The main objective was to investigate the effects of different alkaline medium and surfactant concentrations on the structural, textural, and chemical properties of zeolites with mesopores, as well as to evaluate the role of the surfactant during modification.

2. Materials and Methods

2.1. USY Zeolite Modification

The USY zeolite used as the starting material (code CBV720; SiO_2/Al_2O_3 = 30) was manufactured by Zeolyst (Conshohocken, PA, USA). The molar ratio of the reaction mixture employed in the zeolite modification was 0.0052 HY:x Na2O:100 H2O:y CTAB. The values for the base (x) and surfactant (y) are shown in Table 1.

Table 1. Modification conditions used for preparation of the materials.

Sample	Base Molar Ratio (x)	Surfactant Molar Ratio (y)
YB0-S0.1	0	0.1
YB0.02-S0.1	0.02	0.1
YB0.04-S0.1	0.04	0.1
YB0.06-S0.1	0.06	0.1
YB0.08-S0.1	0.08	0.1
YB0.08-S0	0.08	0
YB0.08-S0.02	0.08	0.02
YB0.08-S0.04	0.08	0.04
YB0.08-S0.06	0.08	0.06
YB0.08-S0.08	0.08	0.08

To prepare the reaction mixture with molar ratio of 0.0052 HY:0.08 Na_2O:100 H_2O:0.1 CTAB, 0.60 g of CTAB was dissolved in 30 mL of aqueous 0.085 mol/L NaOH solution, followed by addition of 1 g USY zeolite under agitation for 20 min. Subsequently, the reaction mixture was subjected to hydrothermal treatment for 20 h at 150 °C in an autoclave. The suspension was filtered and the solid material obtained was washed until pH 7 and dried at 80 °C. Calcination was then performed in a muffle furnace at 550 °C for 8 h, using a heating rate of 2 °C/min. The samples were labeled using the nomenclature YBx-Sy.

2.2. Influence of Surfactant

The zeolites denoted YB0.04-S0-60/6h and YB0.04-S0.08-60/6h were prepared as described in Section 2.1. employing the following molar ratios: 0.0052 HY:0.04 Na_2O:100 H_2O:0 CTAB, and 0.0052 HY:0.04 Na_2O:100 H_2O:0.08 CTAB, respectively. The hydrothermal treatment temperature was 60 °C and the treatment duration was 6 h. A third sample, denoted YB0.04-CTA^+-60/6h, was prepared using a reaction mixture with the molar ratio 0.0052 HY:0.04 Na_2O:100 H_2O:0 CTAB. For the preparation of the sample named YB0.04-CTA^+-60/6h, an ion exchange was first performed with the original zeolite USY by using its mixture with an aqueous solution of CTAB with a concentration of 0.2 mol/L. The ratio used was 1 g zeolite to 100 mL solution. Three consecutive exchanges of 1 h each were performed. At the end of each exchange, the zeolite was washed with distilled water, and after drying it was oven dried at 80 °C. Subsequently, 1 g zeolite containing CTA^+ cations that compensated the negative charges in the structure was dissolved in 30 mL of aqueous 0.0045 mol/L NaOH solution under agitation for 20 min. Afterward, the reaction mixture was submitted to hydrothermal treatment for 6 h at 60 °C in an autoclave. The suspension was filtered and the solid material obtained was washed until pH 7 and dried at 80 °C. Calcination was then performed in a muffle furnace at 550 °C for 8 h using a heating rate of 2 °C/min.

The difference, relative to the zeolites prepared as described above, was that the initial zeolite contained CTA^+ cations that compensated for the negative charges of the zeolite structure. The hydrothermal treatment was performed for 6 h at 60 °C.

2.3. Characterization

X-ray diffractograms were acquired using a Rigaku MiniFlex 600 diffractometer (Tokyo, Japan) operated with Cu Kα radiation (λ = 1.5418 Å). The relative crystallinity (RC) was determined using the ratio of the sums of the peak areas at 23.3°, 26.6°, and 30.9° 2θ for the modified samples, relative to the original sample, as recommended by ASTM. The original zeolite was considered as a standard, with an assumed crystallinity of 100%.

The textural properties of the zeolites were evaluated using nitrogen physisorption isotherms acquired using a Micromeritics ASAP 2020 system. The mesopore diameter distribution was determined by the Barret-Joiner-Halenda (BJH) method, employing the desorption branch of the isotherm. The micropore volume was determined by the t-plot method, using a 0.3–0.5 nm thickness interval. The mesopore volume was determined by the NLDFT method for pores with cylindrical geometry based on the functional density [10,15]. The total quantities of acid sites in the materials were determined by the Ammonia Temperature Programmed Desorption technique (TPD-NH_3), employing a Micromeritics AutoChem II 2920 chemisorption analyzer.

The global silicon/aluminum ratios were determined by energy dispersive spectroscopy (EDS). The analyses were performed using a field emission gun (FEG) electron microscope operated at 20 kV with the samples dispersed on double-sided adhesive carbon tapes.

Quantification of silicon in the filtrate was performed by chemical analysis using an Optima 8000 Inductively Coupled Plasma Optical Emission Spectrometer (ICP-OES).

Analysis of the zeolites by ^{27}Al nuclear magnetic resonance (NMR) was performed using a Bruker Avance III-400 system operated with a 9.4 T magnetic field. The samples were packed into zirconia rotors with external diameter of 4 mm and the measurements were performed at a temperature of 23 °C.

3. Results

3.1. Influence of Surfactant Concentration

The X-ray diffractograms of the modified zeolites and the original USY zeolite (Figure 1) revealed that only the zeolite modified without the addition of a surfactant (YB0.08-S0) did not show diffraction peaks corresponding to the FAU structure.

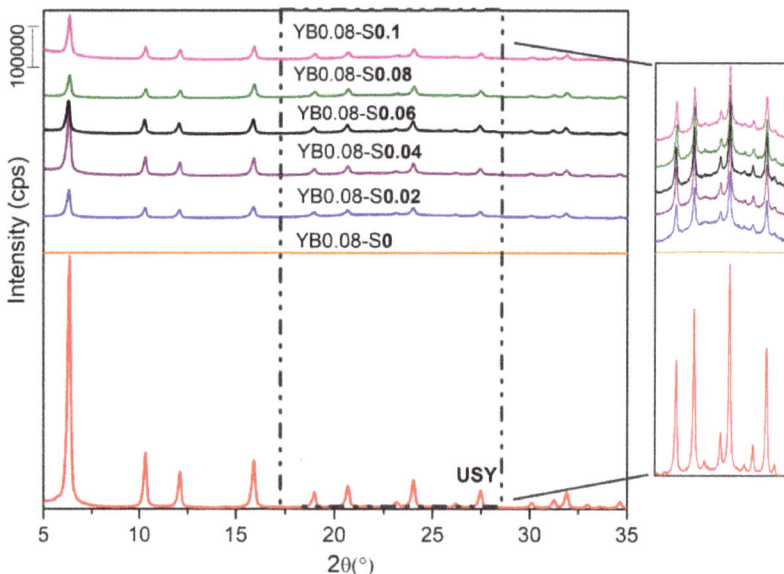

Figure 1. X-ray diffractograms of the samples modified using different concentrations of cetyltrimethylammonium bromide (CTAB).

Calculation of the relative crystallinity (RC) of the modified zeolites (Table 2) revealed an average reduction in RC of approximately 65% under the conditions tested. Notably, the use of CTA$^+$ in the alkaline treatment protected the Si–O–Si bonds from attack by hydroxyl groups (OH$^-$), thus preventing the complete destruction of the zeolite structure. In a study on the desilicalization of zeolite Y with different Si/Al ratios used in this study, the same behavior in the presence of CTAB was verified [14].

Table 2. Relative crystallinity (RC) and textural properties of the modified zeolites, according to the amount of surfactant used during the modification.

Sample	RC (%)	V_{total} [1] (cm^3/g)	V_{micro} [2] (cm^3/g)	V_{meso} [3] (cm^3/g)	S_{ext} [2] (cm^3/g)
USY	100	0.395	0.239	0.115	208
YB0.08-S0	0	0.023	0.002	0.033	20
YB0.08-S0.02	32	0.225	0.087	0.123	165
YB0.08-S0.04	39	0.395	0.107	0.193	364
YB0.08-S0.06	38	0.468	0.091	0.265	536
YB0.08-S0.08	38	0.603	0.092	0.363	597
YB0.08-S0.1	33	0.649	0.077	0.404	661

[1] $P/P_0 = 0.85$; [2] External surface area by t-plot; [3] determined by NLDFT. The pH of the reaction mixture after the USY treatment was about 11.0.

The X-ray diffractograms showed the presence of a halo around 15–30° 2θ (the magnification in Figure 1), possibly due to silicon atoms removed from the structure and deposited in the form of amorphous silica organized by the CTA^+ micelles. For a better visualization of the amorphous halo in the X-ray diffraction (XRD) pattern of sample YB0.08-S0, the graph was plotted on a smaller scale. (Figure S1). At small angles, the X-ray diffractograms of the modified zeolites showed a peak near of 2° 2θ, corresponding to the repetition of crystallographic planes, possibly resulting from the ordering of pores larger than 2 nm, derived from the treatments of the materials (Figure 2).

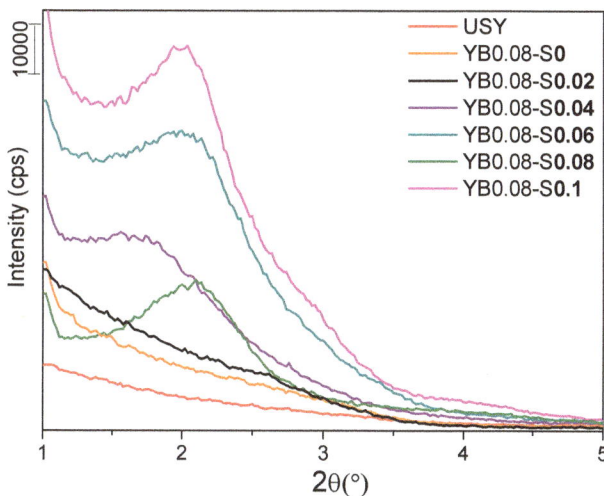

Figure 2. X-ray diffractograms at small angles of the modified zeolites produced using different concentrations of surfactant (CTA^+).

Comparison of the modified zeolites with low (y = 0.02) and high (y = 0.1) surfactant contents showed that the relative crystallinity did not change significantly. In other words, the surfactant content did not considerably influence this parameter.

The initial USY zeolite presented a type I isotherm (Figure 3) according to the IUPAC [16] classification, reflecting a predominance of micropores in its structure. However, the modified materials showed isotherms that were a combination of types I and IV (Figure 3), indicative of the presence of micro- and mesopores in their structures and confirming the effectiveness of the modifications. The porous structure of the YB0.08-S0 zeolite was completely destroyed due to the absence of CTA^+

during the modification, in agreement with the destruction of the faujasite structure shown by the X-ray diffraction analysis (Figure 1).

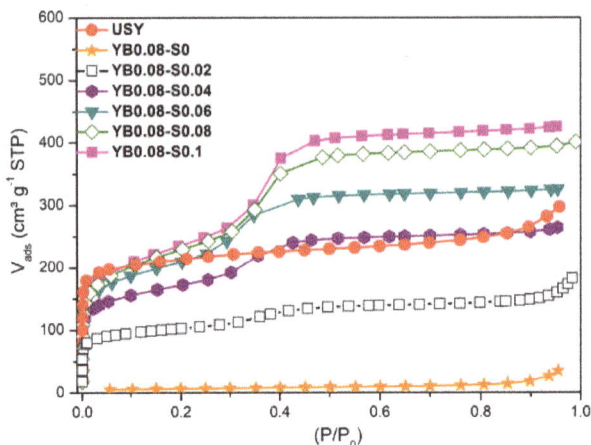

Figure 3. N_2 physisorption isotherms for the original USY zeolite and the modified zeolites produced using different surfactant concentrations.

The larger mesopore volume of the modified samples produced using higher surfactant ratios (Table 2) were probably due to the formation of a higher number of micelles, since these are templates for the formation of mesopores. The external area, determined by the t-plot method, increased with increasing mesopore volume.

Figure 4 shows the volumes of micropores and mesopores as a function of the surfactant ratio used in modification of the original USY zeolite. In the absence of addition of CTA^+, the volume of micro- and mesopores was almost zero because the crystalline structure had been completely destroyed, as shown previously. Addition of the CTA^+ surfactant at different concentrations resulted in an almost constant micropore volume, in agreement with the effect on the relative crystallinity, described previously.

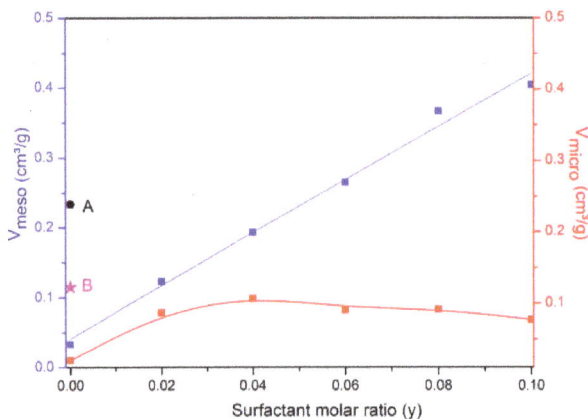

Figure 4. Micropore and mesopore volumes, as a function of surfactant molar ratio. A and B represent the volume of micropores and mesopores in the original USY zeolite, respectively.

The original sample possessed a small quantity of pores with diameters around 3.8 nm, derived from the hydrothermal treatment applied to the Y zeolite to increase its catalytic stability (Figure 5). Modification of the USY zeolite using the lowest surfactant concentration (y = 0.02) maintained part of the original pores but led to the formation of smaller mesopores around 3 nm in size. The use of higher surfactant concentrations led to the disappearance of pores of around 3.8 nm, but increased the amount of mesopores around 3.0 nm in size. Given that the diameters of CTA^+ cation micelles are in the range 3 to 4 nm [7], the formation of mesopores in this same size range could be attributed to the role of the surfactant as a template in the formation of mesopores.

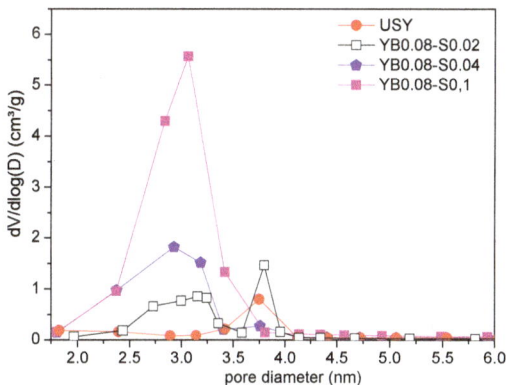

Figure 5. Pore diameter distributions (Barret-Joiner-Halenda method; BJH) of the modified zeolites produced using different CTAB concentrations.

3.2. Influence of Base Concentration

The X-ray diffractograms of the modified USY zeolites presented diffraction peaks characteristic of the faujasite structure, but with lower intensity compared to the diffractogram of the original USY zeolite (Figure 6). The X-ray diffractograms of the modified zeolites produced using higher amounts of base (x = 0.04–0.08) with a more pronounced halo at 2θ of 15–30°, possibly related to amorphous silica, different from the X-ray diffractograms of the modified zeolites produced with lower base content.

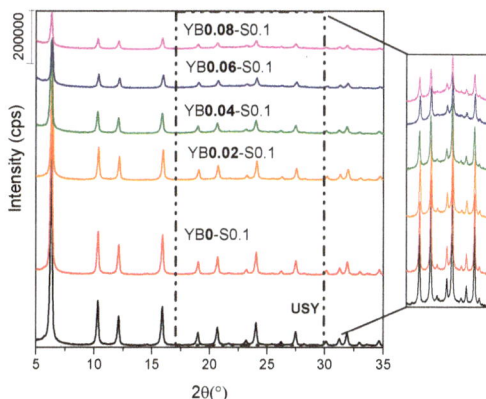

Figure 6. X-ray diffractograms of the modified zeolites produced using different base ratios.

The relative crystallinity of the zeolites decreased as the concentration of the alkaline medium increased (Table 3). This demonstrated the importance of adjusting the concentration of the alkali solution in order to avoid excessive amorphization of the material and conserve the properties of the zeolite.

Table 3. Relative crystallinity and textural properties of the modified zeolites, according to the base concentration used.

Sample	RC (%)	V_{total} [1] (cm³/g)	V_{micro} [2] (cm³/g)	V_{meso} [3] (cm³/g)	S_{ext} [2] (cm³/g)	pH [4]
USY	100	0.395	0.239	0.115	208	-
YB0-S0.1	96	0.396	0.242	0.142	186	3
YB0.02-S0.1	80	0.367	0.199	0.170	209	9
YB0.04-S0.1	52	0.464	0.129	0.229	420	10
YB0.06-S0.1	38	0.627	0.095	0.379	591	11
YB0.08-S0.1	33	0.649	0.077	0.404	661	11

[1] $P/P_0 = 0.85$; [2] t-plot; [3] NLDFT; [4] pH of the reaction mixture after the USY treatment.

The X-ray diffractograms at small angles of the modified zeolites produced using different base concentrations are shown in Figure 7. The YB0-S0.1, YB0.02-S0.1, and original USY zeolites did not present a diffraction peak at 2° 2θ, corresponding to the repetition of the crystallographic planes, probably due to the absence of mesoporosity with a certain degree of organization.

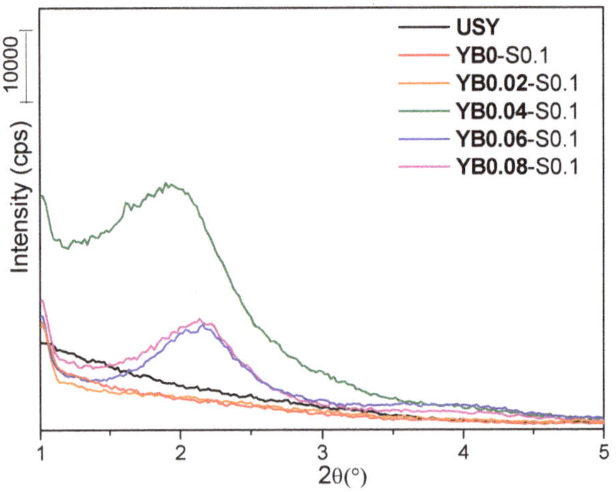

Figure 7. X-ray diffractograms at small angles of the modified zeolites produced using different base concentrations.

The nitrogen adsorption isotherms (Figure 8) revealed the lower nitrogen adsorption by the YB0.02-S0.1 zeolite at a relative pressure of 0.4, indicative of the generation of fewer mesopores due to the milder alkaline treatment. The modified zeolites produced using higher base concentrations presented greater mesopore formation.

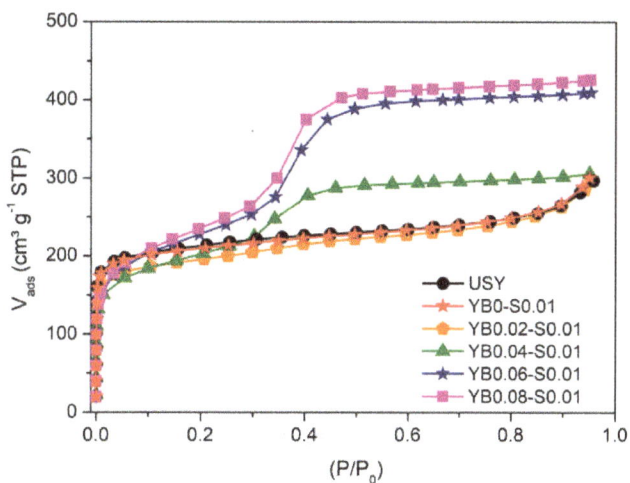

Figure 8. Nitrogen adsorption isotherms of the modified samples produced using different concentrations of the alkaline medium.

From the data presented in Table 3 and Figure 8, no mesoporosity was created when the modification was performed in the absence of a base. This indicated that the formation of SiO^- species by breaking the Si–O–Si bonds due to the action of a base was fundamental for the formation of mesopores.

A positive correlation was found between the NaOH concentration used in the treatment and the mesopore volume of the resulting material. In addition, an increase in the volume of mesopores was accompanied by a decrease in the micropore volume (Table 3, Figure 9), as also found elsewhere [10].

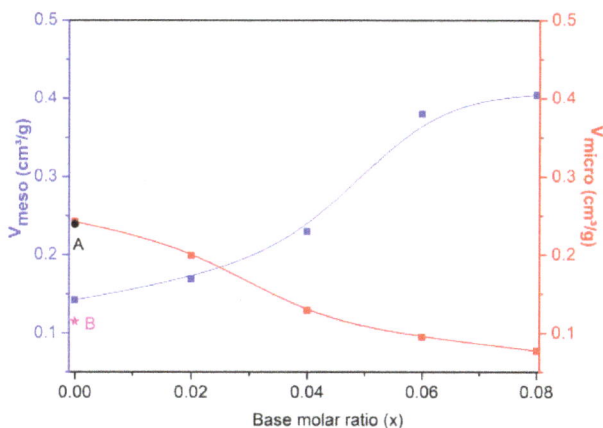

Figure 9. Zeolite mesopore and micropore volumes as a function of the base ratio (x) used for the modification. A and B represent the volume of micropores and mesopores in the original USY zeolite, respectively.

The modified zeolites demonstrated a narrow pore size distribution centered near 3 nm (Figure 10). As the alkaline treatment concentration increased, the volume of the mesopores generated also increased. The mildest alkaline treatment (x = 0.02) was insufficient to create any substantial quantity of mesopores templated by the CTA$^+$ micelles.

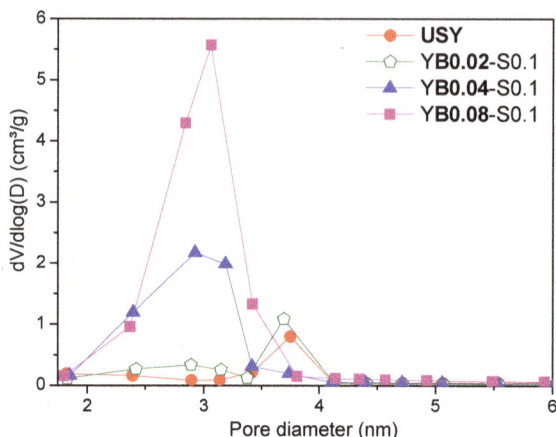

Figure 10. Pore size distributions obtained using the BJH method applied to the isotherms (desorption branch) for the original zeolite and the modified zeolites produced using different base concentrations.

The observed main influence on the modification of the textural properties of the zeolites by the concentration of the alkaline medium was further investigated via chemical analyses of the materials.

The different chemical environments of the aluminum in the zeolites were investigated by ^{27}Al NMR (Figure 11). The signal corresponding to octahedral aluminum (Al^{VI}) was observed at 0 ppm, whereas tetracoordinated aluminum (Al^{IV}) was identified both within and outside the zeolitic structure, as shown by signals located at 60 and 53 ppm, respectively. The spectra indicated that the use of a higher NaOH concentration led to a decrease in the tetracoordinated aluminum in the zeolite structure, accompanied by an increase in non-structural or distorted tetracoordinated aluminum, as reported previously [17].

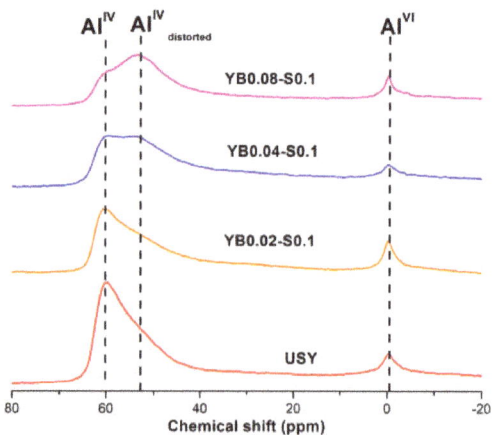

Figure 11. ^{27}Al nuclear magnetic resonance (NMR) spectra of the calcined modified zeolites produced using different alkaline medium concentrations.

This change in the chemical environment of the aluminum led to a decrease in Brønsted acid sites, since these sites are generated due to the presence of tetrahedral aluminum in the zeolite structure. The presence of distorted or external (outside the structure) tetracoordinated aluminum generates Lewis acid sites [17,18].

The TPD-NH$_3$ profiles (Figure 12) showed two peaks: the first around 208–217 °C and the second around 325–360 °C. The peak at the lower temperature could be attributed to the release of weakly adsorbed NH$_3$, whereas the peak at the higher temperature could be explained by the release of NH$_3$ from NH$_4^+$ bound at stronger acid sites [19,20].

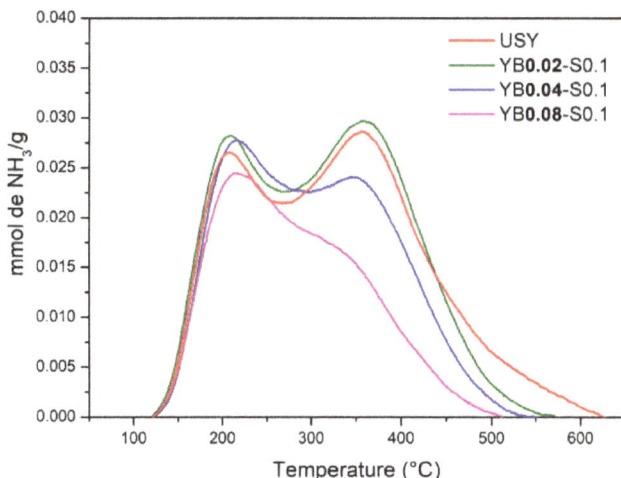

Figure 12. TPD-NH$_3$ profiles of the calcined original zeolite and the zeolites produced using different base concentrations.

Considering the total amounts of ammonia desorbed at higher temperatures (Table 4), the original USY zeolite contained a greater quantity of acid sites compared to the modified zeolites, demonstrating that the process used to create mesoporosity resulted in fewer acid sites [2,10].

Table 4. Quantification of the total acidity of the zeolites.

Sample	T (°C)	NH$_3$ $\left(\frac{\mu mol}{g}\right)$	NH$_{3(total)}$ $\left(\frac{\mu mol}{g}\right)$	Si/Al
USY	209 358	101 446	547	14
YB0.02-S0.1	208 358	144 391	535	15
YB0.04-S0.1	216 350	115 347	462	14
YB0.08-S0.1	216 325	99 245	344	14

A yield of 97.6% was obtained for the modification performed under the most severe conditions (x = 0.08, y = 0.1). ICP analysis showed that the amount of silicon present in the filtrate was insignificant and corresponded to only 0.07% of the silicon present in the original zeolite. This was supported by the constancy of the global Si/Al ratio (Table 4), confirming that the modification led to no significant loss of silicon contained in the material. The Si removed from the structure by the alkaline treatment was subsequently associated with the surfactant micelles and was recrystallized.

3.3. Effect of the Presence of the Surfactant

We evaluated the influence of the presence of the CTA^+ surfactant during modification of the USY zeolite on the generation of mesopores. As observed in Section 3.2., the presence of the surfactant during the alkaline treatment protected the zeolite structure from more intense action of the NaOH. The X-ray diffractograms of the YB0.04-S0.08-60/6h and YB0.04-S0-60/6h zeolites presented characteristic peaks of the faujasite structure (Figure 13).

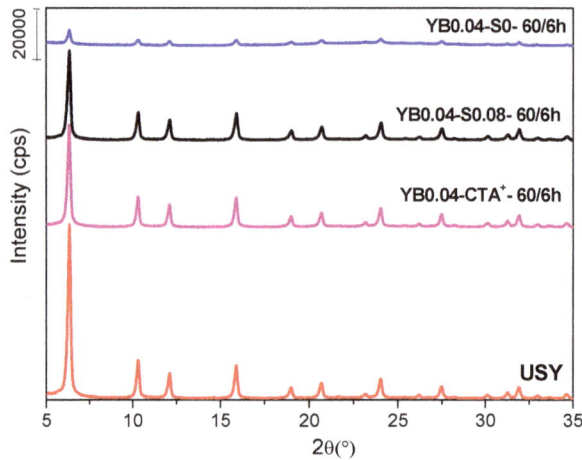

Figure 13. X-ray diffractograms of the zeolites used to study the influence of the presence of CTA^+ cations.

The zeolite modified without the presence of CTA^+ (YB0.04-S0-60/6h) showed a 77% reduction of relative crystallinity, whereas the RC of the zeolite modified with the addition of CTA^+ in the reaction mixture (YB0.04-S0.08-60/6h) decreased by only 6% (Table 5), confirming the protective effect of the surfactant.

Table 5. Relative crystallinity and textural properties of the zeolites used to study the effect of the presence of the CTA^+ cations.

Sample	RC(%)	V_{total} [1] (cm^3/g)	V_{micro} [2] (cm^3/g)	V_{meso} [3] (cm^3/g)	S_{ext} [2] (cm^3/g)
USY	100	0.395	0.239	0.156	208
YB0.04-CTA$^+$-60/6h	98	0.405	0.245	0.160	179
YB0.04-S0.08-60/6h	94	0.426	0.240	0.186	233
YB0.04-S0.-60/6h	23	0.355	0.084	0.271	334

[1] $P/P_0 = 0.85$; [2] t-plot; [3] NLDFT.

The modified USY zeolite produced using CTA^+ cations to compensate for the negative charges of the aluminum (denoted YB0.04-CTA$^+$-60/6h) exhibited high relative crystallinity (Table 5, Figure 13). To verify the presence of the CTA^+ cations in this sample, thermogravimetric analysis (TGA) was performed (Figure S4). The presence of the cations could be explained by the presence of the cetyl ($-C_{16}H_{33}$) and methyl ($-CH_3$) groups of the surfactant, whose occupation of space prevented the hydroxyl (OH^-) groups from disrupting the Si–O–Si bonds protecting the micropores. This explanation was supported by a previous study that reported that the use of tetramethylammonium hydroxide (TMAOH) [14] could protect the zeolite structure from hydroxyl attack, different from the use of the inorganic bases NH_4OH and NaOH.

Analysis of the isotherms (Figure 14) revealed a smaller pore volume of the zeolite modified under mild conditions in the absence of the CTA^+ cation (YB0.04-S0-60/6h) compared to the zeolites modified in the presence of the surfactant. This occurred because the CTA^+ cations were not present to protect the zeolite structure, so the action of the base disrupted a greater quantity of Si–O–Si bonds.

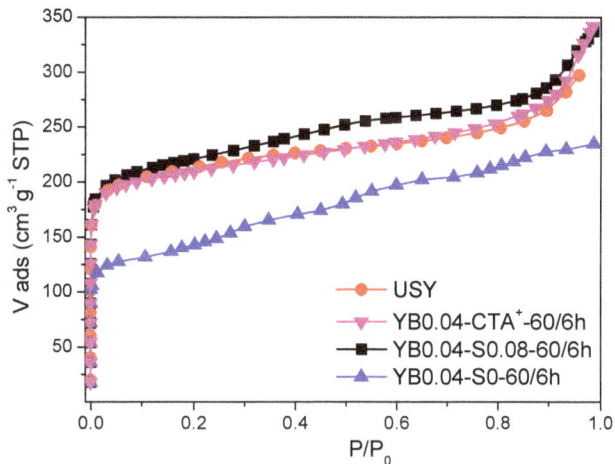

Figure 14. Nitrogen adsorption isotherms for the zeolite samples used to evaluate the effect of the presence of CTA^+.

Analysis of the textural properties of the zeolites (Table 5) showed that the zeolite produced with CTA^+ cations compensating for the charges of the structure (YB0.04-CTA^+-60/6h) did not form mesopores after the modification. This could be explained by the absence of micelle formation during the process, given that no surfactant was added to the reaction mixture.

Analysis of the textural properties of the zeolite modified in the presence of the surfactant under milder hydrothermal treatment conditions (YB0.04-S0.08-60/6h) revealed that the milder alkaline solution and hydrothermal treatment conditions, when used together, were insufficient to form a significant quantity of mesopores. In this case, the micropore and mesopore volumes were virtually identical to those of the original USY zeolite (Table 5).

In the absence of the surfactant, mesopores formed with a concomitant reduction in micropores (Table 5). This likely occurred because the absence of CTA^+ cations facilitated the breaking of the Si–O–Si bonds, with mesopores being formed by desilication, even with the use of mild reaction conditions.

4. Conclusions

The proposed methodology was effective for the creation of mesopores in USY zeolite in the presence of CTA^+ cations. The presence of the CTA^+ cations was fundamental in the modification process, since the cations hindered the attack by hydroxyl (OH^-) groups, avoiding the dissolution of the zeolite crystals during the modification.

The use of an alkaline treatment was essential, since the absence of the base resulted in the lack of mesoporosity creation. Therefore, mesoporosity formation in the USY zeolite required the simultaneous presence of the alkaline medium and the CTA^+ surfactant.

Appl. Sci. **2018**, *8*, 1299

In addition to the protective effect of the CTA$^+$ surfactant, an increase in its concentration had a positive effect on increasing the mesopore volume of the modified zeolites, without reducing the relative crystallinity, different than the behavior observed following variation of the alkaline medium concentration. A higher concentration of the alkaline medium significantly influenced the textural properties and the relative crystallinity values of the modified zeolites.

The total quantities of acid sites of the modified zeolites were lower than for the original USY zeolite. Increasing the alkaline medium concentration used in the modification resulted in lower acidity due to the substantial decrease in the micropore volume. This reduction in the quantity of acid sites could be attributed to changes in the chemical environments of aluminum.

The USY zeolite with Si/Al = 15 was highly sensitive to desilication, preventing the formation of a substantial volume of mesopores without amorphization or even total destruction of the zeolite structure. Therefore, the methodology described here is attractive, as it is a versatile technique that increases the mesopore volume, without risking loss of the zeolitic structure.

Supplementary Materials: The following are available online at http://www.mdpi.com/2076-3417/8/8/1299/s1, Figure S1: X-ray diffractograms of the sample YB0.08-S0, Figure S2: Nitrogen adsorption (closed symbols) and desorption (open desorption) isotherms at 77 K of the modified samples produced using different concentrations surfactant, Figure S3: Nitrogen adsorption (closed symbols) and desorption (open desorption) isotherms at 77 K of the modified samples produced using different concentrations of the alkaline medium, Figure S4: Thermogravimetric analysis of the sample original USY and YB0.04-CTA+-60/6h.

Author Contributions: J.F.S. and D.C. designed the experiments; E.D.F. and J.F.S. performed the experimental work; J.F.S., E.D.F., and D.C. analyzed the data; D.C. provided reagents, materials, and analysis tools; J.F.S. wrote the article, with contributions from D.C. and E.D.F.

Funding: This research received no external funding.

Acknowledgments: The authors are grateful to CAPES/MEC for scholarship support.

Conflicts of Interest: The authors declare no conflicts of interest.

References

1. Giannetto, G. *Zeolitas: Características, Propiedades y Aplicaciones Industriales*, 1st ed.; Editorial Innovación Tecnológica: Caracas, Venezuela, 1990; p. 170.

2. Chal, R.; Cacciaguerra, T.; van Donk, S.; Gérardin, C. Pseudomorphic synthesis of mesoporous zeolite Y. crystals. *Chem. Commun.* **2010**, *46*, 7840–7842. [CrossRef] [PubMed]

3. Pérez-Ramírez, J.; Christensen, C.H.; Egeblad, K.; Christensen, C.H.; Groen, J.C. Hierarchical zeolites: Enhanced utilisation of microporous crystals in catalysis by advances in materials design. *Chem. Soc. Rev.* **2008**, *37*, 2530–2542. [CrossRef] [PubMed]

4. Verboekend, D.; Nuttens, N.; Locus, R.; Van Aelst, J.; Verolme, P.; Groen, J.C.; Pérez-Ramírez, J.; Sels, B.F. Synthesis, characterisation, and catalytic evaluation of hierarchical faujasite zeolites: Milestones, challenges, and future directions. *Chem. Soc. Rev.* **2016**, *45*, 3331–3352. [CrossRef] [PubMed]

5. Li, K.; Valla, J.; Garcia-Martinez, J. Realizing the commercial potential of hierarchical zeolites: New opportunities in catalytic cracking. *Chem. Cat. Chem.* **2014**, *6*, 46–66.

6. Koster, A.J.; Ziese, U.; Verkleij, A.J.; Janssen, A.H.; de Jong, K.P. Three-dimensional transmission electron microscopy: A novel imaging and characterization technique with nanometer scale resolution for Materials Science. *J. Phys. Chem. B* **2000**, *104*, 9368–9370. [CrossRef]

7. Ivanova, I.I.; Kasyanov, I.A.; Maerle, A.A.; Zaikovskii, V.I. Mechanistic study of zeolites recrystallization into micro-mesoporous materials. *Microporous Mesoporous Mater.* **2014**, *189*, 163–172. [CrossRef]

8. García-Martínez, J.; Johnson, M.; Valla, J.; Li, K.; Ying, J.Y. Mesostructured zeolite Y—high hydrothermal stability and superior FCC catalytic performance. *Catal. Sci. Technol.* **2012**, *2*, 987–994. [CrossRef]

9. Ivanova, I.I.; Kuznetsov, A.S.; Ponomareva, O.A.; Yuschenko, V.V.; Knyazeva, E.E. Micro/mesoporous catalysts obtained by recrystallization of mordenite. In *Studies in Surface Science and Catalysis*; Elsevier: Amsterdam, The Netherlands, 2005; pp. 121–128.

10.   Sachse, A.; Grau-Atienza, A.; Jardim, E.O.; Linares, N.; Thommes, M.; García-Martínez, J. Development of intracrystalline mesoporosity in zeolites through surfactant-templating. *Cryst. Growth Des.* **2017**, *17*, 4289–4305. [CrossRef]
11.   Sachse, A.; Wuttke, C.; Díaz, U.; de Souza, M.O. Mesoporous Y zeolite through ionic liquid based surfactant templating. *Microporous Mesoporous Mater.* **2015**, *217*, 81–86. [CrossRef]
12.   Ying, Y.; García-Martínez, J. Mesostructured Zeolitic Materials, and Methods of Making and Using the Same. U.S. Patent 20050239634A1, 27 October 2005.
13.   Verboekend, D.; Vilé, G.; Pérez-Ramírez, J. Mesopore formation in USY and beta zeolites by base leaching: Selection criteria and optimization of pore-directing agents. *Cryst. Growth Des.* **2012**, *12*, 3123–3132. [CrossRef]
14.   Shutkina, O.V.; Knyazeva, E.E.; Ivanova, I.I. Preparation and physicochemical and catalytic properties of micro-mesoporous catalysts based on faujasite. *Pet. Chem.* **2016**, *56*, 138–145. [CrossRef]
15.   Thomas, J.M.; Leary, R.K. A Major Advance in Characterizing Nanoporous Solids Using a Complementary Triad of Existing Techniques Angew. *Chemistry* **2014**, *53*, 12020–12021.
16.   Sing, K.S.W.; Everett, D.H.; Haul, R.A.W.; Moscou, L.; Pierotti, R.A.; Rouquerol, J.; Siemieniewska, T. Reporting physisorption data for gas/solid systems with special reference to the determination of surface area and porosity. *Pure Appl. Chem.* **1985**, *57*, 603–619. [CrossRef]
17.   Van Aelst, J.; Verboekend, D.; Philippaerts, A.; Nuttens, N.; Kurttepeli, M.; Gobechiya, E.; Haouas, M.; Sree, S.P.; Denayer, J.F.M.; Martens, J.A.; et al. Catalyst design by NH_4OH treatment of USY zeolite. *Adv. Funct. Mater.* **2015**, *25*, 7130–7144. [CrossRef]
18.   Peters, A.W.; Wu, C.C. Selectivity effects of a new aluminum species in strongly dealuminated USY containing FCC catalysts. *Catal. Lett.* **1995**, *30*, 171–179. [CrossRef]
19.   Katada, N.; Igi, H.; Kim, J.-H. Determination of the acidic properties of zeolite by theoretical analysis of temperature-programmed desorption of ammonia based on adsorption equilibrium. *J. Phys. Chem. B* **1997**, *101*, 5969–5977. [CrossRef]
20.   Niwa, M.; Katada, N. New method for the temperature-programmed desorption (TPD) of ammonia experiment for characterization of zeolite acidity: A review: TPD of ammonia for characterization of zeolite acidity. *Chem. Rec.* **2013**, *13*, 432–455. [CrossRef] [PubMed]

© 2018 by the authors. Licensee MDPI, Basel, Switzerland. This article is an open access article distributed under the terms and conditions of the Creative Commons Attribution (CC BY) license (http://creativecommons.org/licenses/by/4.0/).

applied
sciences

MDPI

Article

Introduction of Al into the HPM-1 Framework by In Situ Generated Seeds as an Alternative Methodology

Paloma Vinaches [1], Alex Rojas [2], Ana Ellen V. de Alencar [1], Enrique Rodríguez-Castellón [3], Tiago P. Braga [1] and Sibele B. C. Pergher [1,*]

[1] Laboratório de Peneiras Moleculares (LABPEMOL), Instituto de Química, Universidade Federal de Rio Grande do Norte, Campus de Lagoa Nova, 59078-970 Natal, Brazil; palomavinaches@gmail.com (P.V.); ellenvalencar@yahoo.com.br (A.E.V.d.A.); tiagoquimicaufrn@gmail.com (T.P.B.)
[2] Instituto Federal do Maranhão (UFMA), Avenida dos Portugueses, Vila Bacanga, 65080-805 São Luís, Brazil; alex1981rojas@hotmail.com
[3] Departamento de Química Inorgánica, Cristalografía e Mineralogía, Facultad de Ciencias, Universidad de Málaga (UMA), 29071 Málaga, Spain; castellon@uma.es
* Correspondence: sibelepergher@gmail.com; Tel.: +55-84-3342-2323

Received: 10 July 2018; Accepted: 29 July 2018; Published: 13 September 2018

Abstract: An alternative method for the introduction of aluminum into the STW zeolitic framework is presented. HPM-1, a chiral STW zeolite with helical pores, was synthesized in the pure silica form, and an aluminum source was added by in situ generated seeds. Displacements of the peak positions in the Al samples were found in the X-ray diffractograms, indicating the possible incorporation of the heteroatom into the framework. Using an analysis of the ^{29}Si and ^{27}Al magic-angle spinning nuclear magnetic resonance (MAS NMR) spectra, we concluded that the aluminum was effectively introduced into the framework. The ($Al_{TETRAHEDRAL}$/$Al_{OCTAHEDRAL}$) ratio and its textural properties were studied to explain the catalytic ethanol conversion results at medium temperatures. The sample with the lowest Si/Al ratio showed the best results due to its higher surface area and pore volume, in comparison to those observed for the sample with the highest Si/Al ratio, and due to its higher bulk tetrahedral aluminum content, in comparison to the intermediate Si/Al ratio sample. All catalysts were selective to ethylene and diethyl ether, confirming the presence of acidic sites.

Keywords: STW zeolite; aluminosilicate; seeds; 2-ethyl-1,3,4-trimethylimidazolium; hydrofluoric media; ethanol dehydration

1. Introduction

Historically, zeolites have been defined as crystalline porous aluminosilicates, consisting of tetrahedra in which Si^{4+} or Al^{3+} cations are found inside and O^{2-} anions are placed at their vertices [1,2]. Owing to the increasing knowledge of the zeolitic structure, the current definition has evolved to be, crystalline porous materials for which the structures are formed by tetrahedra enclosing a variety of cations in the interior and with O^{2-} anions at the tetrahedral vertices [3].

HPM-1 is a pure Si chiral STW zeolite with helical pores [4]. Its tridimensional channel system can be decomposed into double 4-ring (D4R) units and $[4^6 5^8 8^2 10^2]$ cavities, forming 10-ring helical channels connected by straight 8-ring pores [5,6]. This structure is shown in Figure 1. The zeolite was obtained from the organic structure of the directing agent 2E134TMI (2-ethyl-1,3,4-trimethylimidazolium). These chiral zeolites have become a focus of study for the last few years [7,8]. The importance of synthesizing these structures arises from their application in the petroleum industry and from their use in obtaining chirally pure drugs.

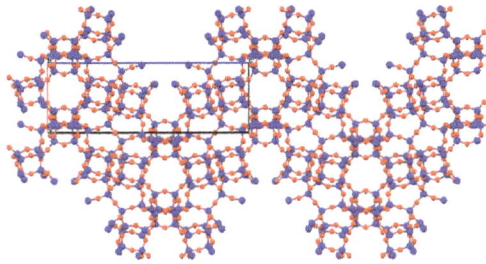

Figure 1. Representation of the STW topology obtained thanks to the .cif from the IZA Database website (with permission from Reference [9], copyright © 2018 Structure Commission of the International Zeolite Association (IZA-SC) [9]) using the Mercury program [10].

Recently, the incorporation of aluminum (Si/Al = 107) in its non-calcined structure was reported, and the obtained structure was named high silica HPM-1 [5]. This material already shows catalytic properties in the isobutene isomerization, which demonstrates its high selectivity. The germanoaluminosilicate was also synthesized by Brand et al. [11] who achieved enantiomeric enrichment. However, a literature search revealed that the calcined aluminosilicate has not been previously described, as only the characterizations of the as-synthesized samples were presented. However, there is interest for understanding its application in catalysis.

Ethanol dehydration is used as a model reaction to confirm the existence of acidic sites [12]. At low reaction temperatures, the main products are ethylene and diethyl ether [13]. The generation of the former product is due to an endothermic reaction (Equation (1)), which is favored thermodynamically at moderate temperatures [14,15].

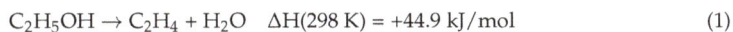

$$C_2H_5OH \rightarrow C_2H_4 + H_2O \quad \Delta H(298\ K) = +44.9\ kJ/mol \tag{1}$$

The second reaction (Equation (2)) is an exothermic reaction resulting in competition between the two products.

$$2C_2H_5OH \rightarrow C_2H_5OC_2H_5 + H_2O \quad \Delta H(298\ K) = -25.1\ kJ/mol \tag{2}$$

As previously mentioned, HPM-1 was obtained as a pure aluminosilicate by the traditional methods of heteroatom introduction in addition to the incorporation of a low quantity of aluminum prior to calcination. Therefore, further study is necessary to understand the behavior of this structure to facilitate the incorporation of different heteroatoms to optimize the synthesis process. For this purpose, the present work aims to study an alternative strategy of Al incorporation and is based on Moura et al.'s published work on the addition of Al into the magadiite framework [16].

2. Materials and Methods

Synthetic procedures: The reagents used in the synthesis were tetraethylorthosilicate (TEOS, 98%, Sigma-Aldrich, St. Louis, MO, USA), aluminum hydroxide (62.23%, Synth, Diadema, SP, Brazil, Al(OH)$_3$,), 2-ethyl-1,3,4-methylimidazolium hydroxide (2E134TMIOH, synthesized), hydrofluoric acid (40%, Sigma-Aldrich, St. Louis, MO, USA), and distilled water.

The cation 2E134TMIOH was synthesized according to the procedure described by Rojas et al. [4]. The synthesis of HPM-1 incorporating aluminum was a variation of the pure silica procedure, and was adapted by the process, reported by Moura et al. [16], to incorporate Al into the lamellar silicate magaadite. In this way, the cation 2E134TMIOH was concentrated up to 1 g/L. Subsequently, TEOS was added to the concentrated cation solution and was left to hydrolyze until the desired 33H$_2$O/SiO$_2$ ratio was reached (due to the ethanol and water evaporation). This procedure was followed according to the weight changes, that is, the weight lost was considered to be ethanol and water evaporation. Once the desired $SiO_2/2E134TMIOH/H_2O$ composition was reached, HF was added and mixed manually for approximately 10–15 min. The gel composition at this point of synthesis was SiO_2: x Al_2O_3: 0.5 SDAOH: 0.5 HF: 4.7 H_2O, where x = 0 (as no aluminum was added yet). Then, the synthesis gel was divided among several Teflon autoclaves that were placed within their respective steel autoclaves. These autoclaves were placed into a rotatory oven at 175 °C for 3 days. At this time, some of the autoclaves were cooled, and the amount of SiO_2 per autoclave was calculated, according to the amount that the synthesis gel weighed. Using the amount of silica per autoclave as a reference, it was possible to calculate the amount of $Al(OH)_3$ needed per autoclave. Further, $Al(OH)_3$ was added to attain the desired compositions (x = 0.015, 0.025, and 0.035). After this, the autoclaves were placed in rotation again at 175 °C for 1 more day. The products were then filtered under a vacuum and washed with an abundance of distilled water. The pH of the initial gel, at the time of aluminum introduction, was 7, and the pH at the end of the synthesis was 8, but it decreased again to 7 when the product was washed. Finally, the samples were calcined at 550 °C for 6 h. This process is summarized in Figure S1 (Supplementary Material).

Characterization: The samples were analyzed by XRD (Bruker D2-Phaser, Bruker, Billerica, MA, USA, with a Lynxeye detector and Cu radiation, using a divergent slit of 0.6 mm, a central slit of 1 mm, a measuring step of 0.004°, and an acquisition time of 0.5 s) to identify the compounds and their crystalline structure. Calibration of the instrument was performed using a National Institute of Standards and Technology (NIST) corundum standard for every measurement in order to ensure that all the samples were comparable. The samples were also analyzed by X-ray fluorescence (XRF, EDX-720/800 HS, Shimadzu, Kioto, Japan) to calculate the total Si/Al ratio (including Al_{TETRA} and Al_{OCTA}). Magic-angle spinning solid-state nuclear magnetic spectroscopy (MAS NMR) was used to calculate the actual Si/Al_{TETRA} ratio of the framework and the Al_{TETRA}/Al_{OCTA} ratio of the samples. The ^{29}Si and ^{27}Al MAS NMR spectra were recorded using a Bruker AVIII HD 600 NMR spectrometer (Bruker, Billerica, MA, USA) (with a field strength of 14.1 T) at 156.4 MHz with a 2.5 mm triple-resonance DVT probe that used zirconia rotors at the spinning rates of 15 kHz (^{29}Si) and 20 kHz (^{27}Al). The ^{29}Si experiments were performed with proton decoupling (cw (continuous wave) sequence) by applying a single pulse ($\pi/2$), an excitation pulse of 5 μs, and a 60 s relaxation delay to obtain 10,800 scans. The ^{27}Al experiments were also performed with proton decoupling (cw sequence) by applying a single pulse ($\pi/12$), an excitation pulse of 1 μs, and a 5 s relaxation delay to obtain 200 scans. The chemical shifts were referenced to as an external solution of tetramethylsilane (TMS) and to an external solution of 1 M of Al $(NO_3)_3$ for ^{29}Si and ^{27}Al, respectively. Field Emission Gun Scanning Electron Microscopy (MIRA3 FEG-SEM, Tescan, Brno, Czech Republic) was used to study the morphology of the samples synthesized with aluminum. The samples were studied by XPS (X-ray photoelectron spectroscopy, Physical Electronics PHI-750 spectrometer, Physical Electronics, Chanhassen, MN, USA) with an X-ray radiation source of Mg Kα (1253.6 eV) and referenced to as C *1s* (284.8 eV) to characterize the Si/Al_{TETRA} ratio and the Al state at the surface. Finally, to study the textural properties of the samples, the calcined samples were pretreated at 200 °C under a vacuum overnight in a Micromeritics Asap 2020 (Micromeritics, Norcross, GA, USA) and were examined using nitrogen as a probe molecule.

Ethanol dehydration: The ethanol used in this reaction was obtained from Sigma-Aldrich.

An amount of 0.1 g of each catalyst was activated at 350 °C and at atmospheric pressure for an hour in an N_2 atmosphere. The catalytic tests were performed in a fixed-bed flow reactor at atmospheric pressure and at 250 °C. A mixture containing N_2 and ethanol vapor (25 mL/min) was stabilized using drag gas (N_2) through a steam saturator system containing ethanol at 25 °C. The outlet gases were analyzed by gas chromatography (GC, Clarus 680, Perkin Elmer, Waltham, MA, USA) equipped with a flame ionization chamber (FIC) and a capillary column.

The ethanol conversion [12] is defined by Equation (3):

$$\%X(EtOH) = (N(EtOH_{in}) - N(EtOH_{out}))/(N(EtOH_{in})) \times 100 \tag{3}$$

For each product, the selectivity [12] can be calculated using Equation (4):

$$\%S(Product\ i) = N(Product\ i)/(\sum_{0}^{i}N(Product\ i)) \times 100 \tag{4}$$

3. Results and Discussion

The samples were obtained by hydrothermal synthesis after four days at 175 °C. The aluminum was added after the third day of synthesis, and the samples were then returned to the oven for the fourth day of heating. Afterwards, the samples were calcined at 550 °C. The complete synthesis is described in the Materials and Methods section, and the Si/Al ratios of the products are summarized in Table 1. The total Si/Al ratio of the products was initially measured by the XRF technique, and the results were similar to the Si/Al ratio of the synthesis gel.

Table 1. Obtained products.

Sample	Si/Al Synthesis Gel (Calculated)	Si/Al Products (XRF)
SiAl15	15	19.4
SiAl25	25	25.4
SiAl35	35	38.0
Pure_Silica	∞	∞

X-ray diffraction (XRD) experiments of the calcined samples were performed to prove that HPM-1 was obtained (Figures 2 and 3). To improve the visibility, the diffractograms were slightly shifted in their intensity. The Pure_Silica sample demonstrates the reproducibility of the original synthesis [4] and is used as the reference for comparison to the other obtained samples.

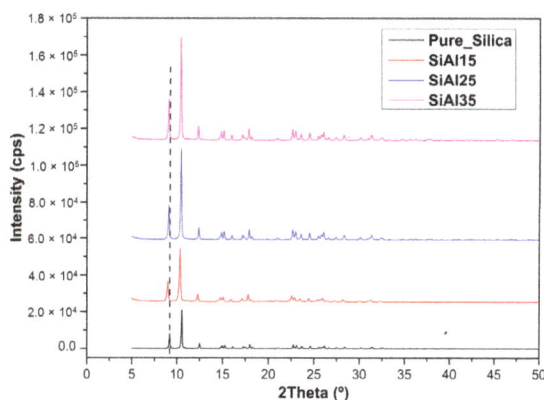

Figure 2. X-ray diffractograms of the calcined products.

It is possible to compare the peak positions of the samples as an indication of the incorporation of a heteroatom in the framework [17]. The substitution of an atom by a different atom in the unit cell results in a distortion of the cell parameters, owing to its different characteristics (e.g., volume, charge). Since the Bragg reflections of an X-ray diffractogram correspond to the different planes and atomic positions, a slight change in these positions, or atomic substitutions, will be reflected in the displacement of the 2θ values of the peaks. In this case, we are studying the incorporation of aluminum, so a shift to lower 2θ values, relative to those of the pure silica samples, indicates that aluminum was effectively incorporated into the framework. Figure 3 shows the 2θ region between 8° and 13°, confirming that this displacement did occur in the three samples when Al was added, possibly suggesting that the aluminum entered the zeolitic framework. As expected, the SiAl15 sample with the highest quantity of Al added showed the largest shift in the diffractogram. Nevertheless, this observation still needs to be confirmed by other techniques because it is difficult to quantify the Si/Al$_{TETRA}$ ratio with the X-ray diffractograms alone.

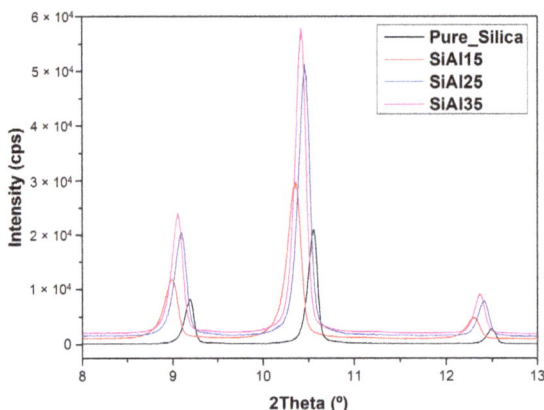

Figure 3. Magnified view of the X-ray diffractograms of the calcined samples.

The difference in the intensities of the same peak positions, among the samples, was another interesting observation from the X-ray diffractograms. These intensity values were normalized to create the average relative crystallinity index, with the number one assigned to the highest intensity of the four peaks at the same 2θ range. To obtain the average, the peak intensities were measured in the following 2θ ranges, 10.3–10.5°, 12.3–13.5°, 16.6–16.3°, and 24.1–24.8°. The results are shown in Table 2. Comparing the aluminosilicate samples, a lower quantity of aluminum in the synthesis helped to obtain a higher average crystallinity index. Surprisingly, the Pure_Silica sample showed the opposite behavior, probably because the presence of aluminum influenced its nucleation and the growth procedure was accelerating it. These results do not mean that the samples are amorphous; this is just a comparison among them as previously reported in the literature [18].

Table 2. Relative crystallinity index.

Sample	Average Relative Crystallinity Index
SiAl15	0.62
SiAl25	0.97
SiAl35	1.00
Pure_Silica	0.34

Studying the magic-angle spinning (MAS) NMR spectra of the calcined solids can provide more precise information on the Si and Al states, since this technique can calculate the Si/Al ratio, assuming that it is larger than seven (proven later). Therefore, the samples synthesized with aluminum were analyzed with this technique, as described by Pace [12]. Two different regions were observed in the ^{29}Si MAS NMR spectra of the calcined samples, which were synthesized with aluminum (Figure 4a, Table 3). This includes the chemical environment, at approximately 105 ppm to 120 ppm, in which all Si atoms, surrounded by O atoms, were bonded to four additional Si atoms. Additionally, the second region centered at approximately 102 ppm which corresponds to the Si[OSi]$_3$[OAl] species, that is, with one Al atom replacing one of the Si atoms in the second coordination sphere [4,12]. The bands of the first region were assigned to Si [OSi]$_4$ outside of the D4R cages (at approximately −115.5 ppm) and to Si [OSi]$_4$ in the D4R cages (at approximately 109 ppm) [19]. Comparing these chemical shifts with those reported for pure silica (published in [4,19]), a slight displacement, which is typical of the heteroatom introduction in a pure silica zeolite, was observed. In addition, the Si[OSi]$_3$[OAl] species in the pure silica sample, confirming the introduction of aluminum in the framework, was not identified. This absence was reported previously for the pure silica form synthesized that employed HF and is explained as being due to the lack of connectivity defects after calcination [19,20]. The sample with the highest Si/Al$_{TETRA}$ ratio corresponded to the sample synthesized with the highest (Si/Al)$_{GEL}$ ratio.

The ^{27}Al MAS NMR spectra (Figure 4b, Table 3) of the calcined samples also confirmed the existence of the aluminum framework, corresponding to the tetrahedral aluminum represented by the band at approximately 60 ppm. A band at approximately 7 ppm, related to octahedral aluminum, also appeared, with a variation in the intensity among the samples. These results showed that a higher aluminum concentration in the synthesis gel led to a lower content of bulk octahedral aluminum.

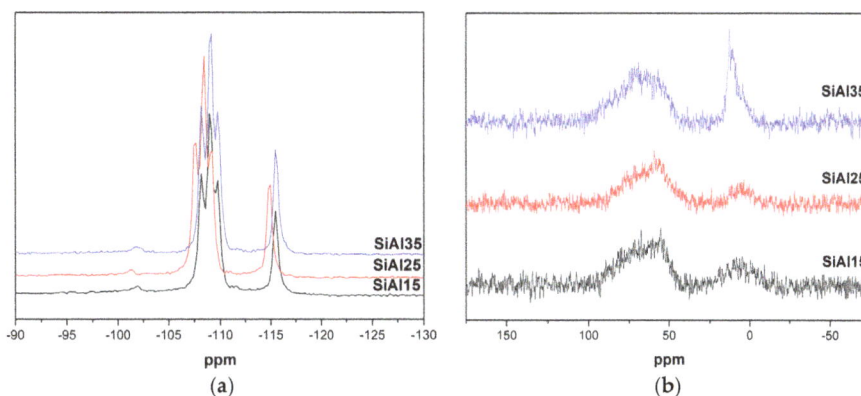

Figure 4. ^{29}Si (**a**) and ^{27}Al (**b**) MAS NMR (Magic-angle spinning solid-state nuclear magnetic spectroscopy) spectra of the calcined samples.

Table 3. ^{29}Si MAS NMR (Magic-angle spinning solid-state nuclear magnetic spectroscopy) chemical shifts of the calcined samples SiAl15, SiAl25, and SiAl35.

Sample	Si/Al$_{TETRA}$		Al$_{TETRA}$/Al$_{OCTA}$	
	MAS NMR	**XPS**	**MAS NMR**	**XPS***
SiAl15	185.0	7.7	1.6	0.6
SiAl25	192.6	11.1	1.3	1.1
SiAl35	192.3	11.9	0.9	1.1

The three [Al] HPM-1 samples were also studied by X-ray photoelectron spectroscopy (XPS) because this technique provides fundamental information regarding the surface chemistry for catalytic reactions [21]. This technique had never previously been reported to study the aluminum distribution in HPM-1, as far as we know. Several interesting differences were found among the samples. Figure 5 presents the Al 2*p* signals from each sample, showing differences in the binding energy (BE, eV) values among the samples. These variations were in the range of the beta zeolite (differences of about ±0.5 eV in BE), a topology that also presents a chiral polymorph [22]. The different (Si/Al$_{TETRA}$)$_{SURFACE}$ ratios were calculated from these spectra and the Si 2*p* spectrum (Table 3). The results followed the same tendency as the ^{29}Si MAS NMR spectra, and showed that the surface was richer than the bulk in Al, resulting in a higher amount of Al incorporated in the external layer of the zeolite. This measurement could also be taking the octahedral aluminum into account. Therefore, the $(Al_{TETRA} / Al_{OCTA})_{SURFACE}$ ratio was calculated, and the modified Auger parameter of Al (α') [23], from the different samples, was obtained according to Equation (5):

$$\alpha' = 1253.6 + KE(Al_{KLL}) - KE(Al\ 2p) \tag{5}$$

where $KE(Al_{KLL})$ is the kinetic energy of the Auger electron at Al_{KLL} (Figure 6), and $KE(Al\ 2p)$ is the kinetic energy of the Al 2p photoelectron. A comparison of these results with the results of ^{27}Al MAS NMR shows that, for SiAl25 and SiAl35, it is similarly indicated that the tetrahedral aluminum would be easily accessible as acidic sites. In the case of SiAl15, the octahedral aluminum content may interfere with the accessibility of the tetrahedral aluminum in future reactions. Despite the increasing ratio from SiAl15 to SiAl35, the results were as expected. The octahedral aluminum was outside of the framework; it was placed on the surface. The data obtained from the XPS analysis are summarized in Tables 3 and 4.

Table 4. XPS (X-ray photoelectron spectroscopy) results of the calcined samples.

Sample	Binding Energy Si 2p (eV)	Binding Energy Al 2p (eV)	α' Al$_{TETRA}$ (eV)	α' Al$_{OCTA}$ (eV)
SiAl15	102.6	74.5	1458.4	1460.8
SiAl25	103.7	75.0	1457.2	1460.2
SiAl35	102.7	74.0	1458.4	1459.8

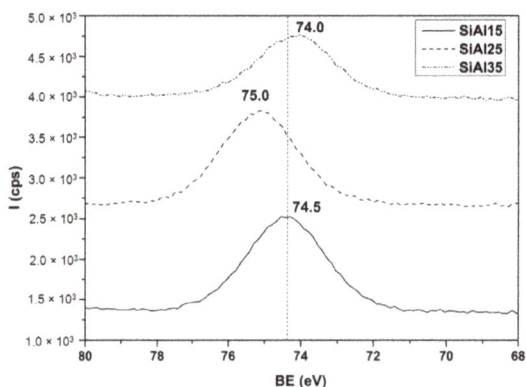

Figure 5. Al 2p signals of the calcined samples (XPS spectra).

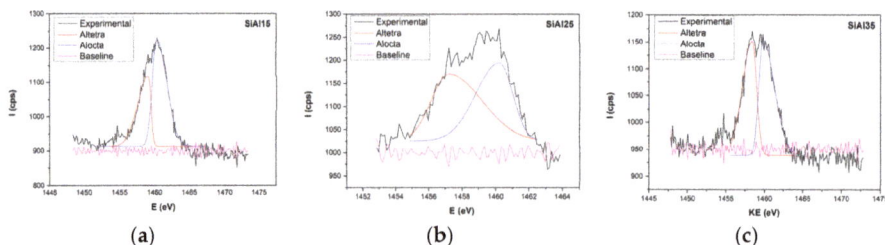

Figure 6. Al$_{KLL}$ signals of the calcined samples (XPS spectra): (**a**) SiAl15; (**b**) SiAl25, and (**c**) SiAl35.

The SiAl25 sample was observed using field emission gun scanning electron microscopy (FEG-SEM), the obtained micrograph is shown in Figure 7, and the Supplementary Material in Figure S2.

The resulting morphology is similar to a rice grain, which has the same reported morphology for the pure silica HPM-1 [4].

Finally, to determine the acidic or basic character of the zeolite, the ethanol dehydration model reaction was used [12]. Figure 8 and Table 5 show the catalytic performance of the [Al] HPM-1 samples. Comparing all the samples, a greater ethanol conversion value was observed for the SiAl25 sample, and lower conversions were observed for the SiAl15 and SiAl35 samples. This order of conversion performance was explained by several factors. The first factor is the BET (Brunauer-Emmett-Teller) area (SBET, Table 5, Supplementary Material Figures S3–S5, and Tables S1–S3), which is directly related to the highest conversion values, because the highest SBET allows a higher quantity of accessible acidic sites [12]. Therefore, SiAl25 and SiAl35 were expected to show higher conversion values than SiAl15. The second factor influencing the ethanol conversion was the amount of octahedral aluminum. While SiAl15 had the highest Si/Al_{TETRA} ratio, its $(Al_{TETRA}/Al_{OCTA})_{XPS}$ ratio was low, implying that the total pore volume (TPV, Table 5, Supplementary Material Tables S1–S3) was lower than in the other cases, thus preventing interaction between the ethanol and the active sites and making the diffusion of the molecules in the pores more difficult. The difference in the conversion performance between SiAl25 and SiAl35 was due to the bulk tetrahedral aluminum content, which was greater for SiAl25 than for SiAl35. Finally, even though SiAl35 has double the area of SiAl15, they presented a similar conversion value, probably because the higher area value was in consideration of the area of the octahedral aluminum, and did not interfere with the reaction. The Pure_Silica sample was also tested for the dehydration of ethanol and showed a very low activity (approximately 2–12%), which is practically inactive, owing to the minimal amount of Si-OH sites reported in a fluoride medium [4,12,19,20].

It is well known that the product selectivity in ethanol conversion is directly related to the acid strength of the material [24]. For the study of sample selectivity at 250 °C, ethanol was catalytically converted via the intramolecular and intermolecular dehydration reactions [25] to produce ethylene and diethylether (DEE), respectively (Table 5, Figure 8)—products that are only formed at acidic sites. This result demonstrated the high acidic character of the samples; the SiAl25 sample was the most acidic sample. Comparing the three cases, at a high temperature and conversion, DEE is produced with high selectivity, as expected, due to the exothermic nature of the reaction [14,15]. The differences in the observed selectivities were small and probably related to the quantity of the available acidic sites because the selectivity results appeared to follow the same pattern.

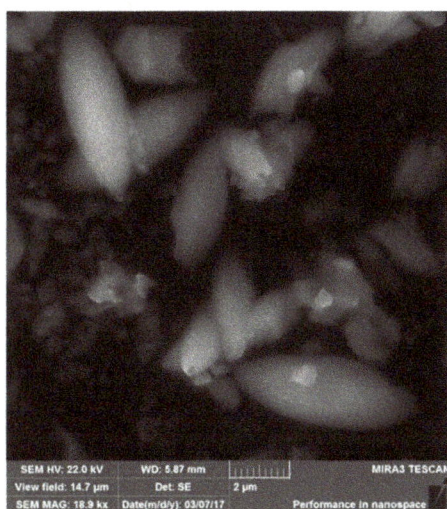

Figure 7. Micrograph of calcined sample SiAl25.

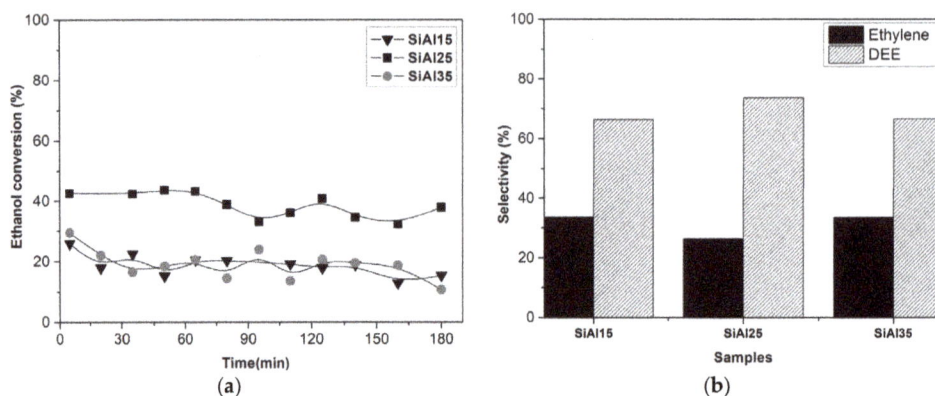

Figure 8. (**a**) Ethanol conversion as a function of time at 250 °C of the calcined [Al] HPM-1 samples and (**b**) product selectivity.

Table 5. Textural and catalysis results of the calcined samples SiAl15, SiAl25, and SiAl35.

Sample	Conversion (Media, %)	Selectivity—Ethylene (%)	Selectivity—DEE (%)	SBET (m² g⁻¹)	TPV * (cm³ g⁻¹)
SiAl15	20	33.6	66.4	332	0.19
SiAl25	38	26.3	73.7	631	0.26
SiAl35	20	33.4	66.6	623	0.26

* TPV (total pore volume) included the mesopore volume due to the space between particles. DEE: Diethylether. SBET: Brunauer-Emmett-Teller area.

4. Conclusions

This work studied a different method for the introduction of Al in the HPM-1 framework using in situ generated seeds. Three samples of [Al] HPM-1 were studied after calcination: SiAl15, SiAl25, and SiAl35. In every case, the SiAl15 product had the lowest Si/Al_{TETRA} ratio, but also had the lowest average crystallinity index and the lowest $(Al_{TETRA}/Al_{OCTA})_{XPS}$ of the samples with aluminum. SiAl25 and SiAl35 had similar Si/Al_{TETRA} ratios, even though SiAl35 had more bulk octahedral aluminum. All samples were active for ethanol dehydration, with SiAl25 being the most active because it had the highest values of S_{BET}, the highest total pore volume, and a higher bulk tetrahedral aluminum content than that of SiAl35. The three catalysts were active to ethanol dehydration, therefore implying that they contained acidic sites. The selectivity results also demonstrated that ethylene production manifests the influence of the Al_{TETRA} acidic sites.

Supplementary Materials: The following are available online at http://www.mdpi.com/2076-3417/8/9/1634/s1, Figure S1: Scheme of the synthesis, Figure S2: SEM/EDX mapping of the sample SiAl25, Figure S3: Nitrogen sorption isotherm (SiAl15), Table S1: Textural analysis of the calcined sample SiAl15, Figure S4: Nitrogen sorption isotherm (SiAl25), Table S2: Textural analysis of the calcined sample SiAl25, Figure S5: Nitrogen sorption isotherm (SiAl35), Table S3: Textural analysis of the calcined sample SiAl35.

Author Contributions: Conceptualization and Methodology, P.V., A.R. and S.B.C.P.; Synthesis, P.V. and A.R.; Characterization, P.V., A.R., E.R.-C. and S.B.C.P.; Catalysis, P.V., A.E.V.d.A. and T.P.B.; Writing, P.V., A.R., A.E.V.d.A., E.R.-C., T.P.B. and S.B.C.P.

Funding: This research was funded by CAPES (Brazil), MINECO (Spain) project number CTQ2015-68951-C3-3-R, and FEDER funds.

Acknowledgments: The authors acknowledge PPGCEM, UFRN, UERN and UMA for the installations and financial support. The authors are also grateful to LABPEMOL (Laboratório de Peneiras Moleculares, UFRN) for the installations for the synthesis, catalysis and the XRD technique, UMA for the MAS NMR and XPS techniques, UERN Chemistry Department for the FEG-SEM micrographs, XRF-DEMAT (Laboratório de Caracterização Estrutural de Materiais, Departamento de Engenharia de Materiais, UFRN) for the XRF data, and the Institute of Chemistry (UFRN) for the IR technique.

Conflicts of Interest: The authors declare no conflict of interest. The funders had no role in the design of the study; in the collection, analyses, or interpretation of data; in the writing of the manuscript, and in the decision to publish the results.

References

1. Flanigen, E.M. Chapter 2 zeolites and molecular sieves an historical perspective. In *Studies in Surface Science and Catalysis*; van Bekkum, H., Flanigen, E.M., Jansen, J.C., Eds.; Elsevier: Amsterdam, The Netherlands, 1991; Volume 58, pp. 13–34.
2. Mintova, S.; Barrier, N. *Verified Synthesis of Zeolitic Materials*, 3rd ed.; Elsevier (On behalf of the Synthesis Commission of the International Zeolite Association): Amsterdam, The Netherlands, 2016; p. 405.
3. Camblor, M.A.; Hong, S.B. Synthetic silicate zeolites: Diverse materials accessible through geoinspiration. In *Porous Materials*; Bruce, D.W., O'Hare, D., Walton, R.I., Eds.; John Wiley & Sons, Ltd.: Hoboken, NJ, USA, 2010.
4. Rojas, A.; Arteaga, O.; Kahr, B.; Camblor, M.A. Synthesis, structure, and optical activity of HPM-1, a pure silica chiral zeolite. *J. Am. Chem. Soc.* **2013**, *135*, 11975–11984. [CrossRef] [PubMed]
5. Jo, D.; Hong, S.B.; Camblor, M.A. Monomolecular skeletal isomerization of 1-butene over selective zeolite catalysts. *ACS Catal.* **2015**, *5*, 2270–2274. [CrossRef]
6. Jo, D.; Lee, K.; Park, G.T.; Hong, S.B. Acid site density effects in zeolite-catalyzed 1-butene skeletal isomerization. *J. Catal.* **2016**, *335*, 58–61. [CrossRef]
7. Davis, M.E. Zeolites from a materials chemistry perspective. *Chem. Mater.* **2014**, *26*, 239–245. [CrossRef]
8. Bueno-Perez, R.; Balestra, S.R.G.; Camblor, M.A.; Min, J.G.; Hong, S.B.; Merkling, P.J.; Calero, S. Influence of flexibility on the separation of chiral isomers in STW-type zeolite. *Chem. A Eur. J.* **2018**, *24*, 4121–4132. [CrossRef] [PubMed]
9. Baerlocher, C.; McCusker, L.B. International Zeolite Association. Database of Zeolite Structures. 2007. Available online: http://www.iza-structure.org/databases/ (accessed on 23 July 2018).
10. Macrae, C.F.; Edgington, P.R.; McCabe, P.; Pidcock, E.; Shields, G.P.; Taylor, R.; Towler, M.; van de Streek, J. Mercury: Visualization and analysis of crystal structures. *J. Appl. Cryst.* **2006**, *39*, 453–457. [CrossRef]
11. Brand, S.K.; Schmidt, J.E.; Deem, M.W.; Daeyaert, F.; Ma, Y.; Terasaki, O.; Orazov, M.; Davis, M.E. Enantiomerically enriched, polycrystalline molecular sieves. *Proc. Natl. Acad. Sci. USA* **2017**, *114*. [CrossRef] [PubMed]
12. Pace, G.G. *Zeolitas: Características, Propiedades y Aplicaciones Industriales*; Editorial Innovacín Tecnológica, Facultad de Ingeniería, UCV: Caracas, Venezuela, 2000.
13. Xin, H.; Li, X.; Fang, Y.; Yi, X.; Hu, W.; Chu, Y.; Zhang, F.; Zheng, A.; Zhang, H.; Li, X. Catalytic dehydration of ethanol over post-treated ZSM-5 zeolites. *J. Catal.* **2014**, *312*, 204–215. [CrossRef]
14. Phung, T.K.; Busca, G. Diethyl ether cracking and ethanol dehydration: Acid catalysis and reaction paths. *Chem. Eng. J.* **2015**, *272*, 92–101. [CrossRef]
15. Phung, T.K.; Busca, G. Ethanol dehydration on silica-aluminas: Active sites and ethylene/diethyl ether selectivities. *Catal. Commun.* **2015**, *68*, 110–115. [CrossRef]
16. Moura, H.M.; Bonk, F.A.; Vinhas, R.C.G.; Landers, R.; Pastore, H.O. Aluminium-magadiite: From crystallization studies to a multifunctional material. *CrystEngComm* **2011**, *13*, 5428–5438. [CrossRef]
17. Vinaches, P.; Rebitski, E.P.; Alves, J.A.B.L.R.; Melo, D.M.A.; Pergher, S.B.C. Unconventional silica source employment in zeolite synthesis: Raw powder glass in mfi synthesis case study. *Mater. Lett.* **2015**, *159*, 233–236. [CrossRef]
18. Vinaches, P.; Pergher, S.B.C. Organic structure-directing agents in sapo synthesis: The case of 2-ethyl-1,3,4-trimethylimidazolium. *Eur. J. Inorg. Chem.* **2018**, *2018*, 123–130. [CrossRef]
19. Rojas, A.; Camblor, M.A. A pure silica chiral polymorph with helical pores. *Angew. Chem. Int. Ed.* **2012**, *51*, 3854–3856. [CrossRef] [PubMed]

20. Villaescusa, L.A.; Márquez, F.M.; Zicovich-Wilson, C.M.; Camblor, M.A. Infrared investigation of fluoride occluded in double four-member rings in zeolites. *J. Phys. Chem. B* **2002**, *106*, 2796–2800. [CrossRef]

21. Bare, S.R.; Knop-Gericke, A.; Teschner, D.; Hävacker, M.; Blume, R.; Rocha, T.; Schlögl, R.; Chan, A.S.Y.; Blackwell, N.; Charochak, M.E.; et al. Surface analysis of zeolites: An XPS, variable kinetic energy XPS, and low energy ion scattering study. *Surf. Sci.* **2016**, *648*, 376–382. [CrossRef]

22. Collignon, F.; Jacobs, P.A.; Grobet, P.; Poncelet, G. Investigation of the coordination state of aluminum in β zeolites by X-ray photoelectron spectroscopy. *J. Phys. Chem. B* **2001**, *105*, 6812–6816. [CrossRef]

23. Gómez-Cazalilla, M.; Mérida-Robles, J.M.; Gurbani, A.; Rodríguez-Castellón, E.; Jiménez-López, A. Characterization and acidic properties of Al-SBA-15 materials prepared by post-synthesis alumination of a low-cost ordered mesoporous silica. *J. Solid State Chem.* **2007**, *180*, 1130–1140. [CrossRef]

24. Martins, L.; Cardoso, D.; Hammer, P.; Garetto, T.; Pulcinelli, S.H.; Santilli, C.V. Efficiency of ethanol conversion induced by controlled modification of pore structure and acidic properties of alumina catalysts. *Appl. Catal. A Gen.* **2011**, *398*, 59–65. [CrossRef]

25. Santos, R.C.R.; Pinheiro, A.N.; Leite, E.R.; Freire, V.N.; Longhinotti, E.; Valentini, A. Simple synthesis of Al_2O_3 sphere composite from hybrid process with improved thermal stability for catalytic applications. *Mater. Chem. Phys.* **2015**, *160*, 119–130. [CrossRef]

© 2018 by the authors. Licensee MDPI, Basel, Switzerland. This article is an open access article distributed under the terms and conditions of the Creative Commons Attribution (CC BY) license (http://creativecommons.org/licenses/by/4.0/).

applied
sciences

MDPI

Article

Synthesis of Zeolite A from Metakaolin and Its Application in the Adsorption of Cationic Dyes

Priscila Martins Pereira [1], Breno Freitas Ferreira [1], Nathalia Paula Oliveira [1], Eduardo José Nassar [1], Katia Jorge Ciuffi [1], Miguel Angel Vicente [2,*], Raquel Trujillano [2], Vicente Rives [2], Antonio Gil [3], Sophia Korili [3] and Emerson Henrique de Faria [1,*]

[1] Grupo de Pesquisas em Materiais Lamelares Híbridos (GPMatLam), Universidade de Franca, Av. Dr. Armando Salles Oliveira, CEP 14404-600, Franca 201, Brazil; priscilapereira@hotmail.com (P.M.P.); brenofreitasferreira@gmail.com (B.F.F.); nathpaulaoliveira@gmail.com (N.P.O.); ejnassar@unifran.br (E.J.N.); katia.ciuffi@unifran.edu.br (K.J.C.)

[2] GIR-QUESCAT-Departamento de Química Inorgánica, Universidad de Salamanca, 37008 Salamanca, Spain; rakel@usal.es (R.T.); vrives@usal.es (V.R.)

[3] INAMAT-Departamento de Química Aplicada, Universidad Pública de Navarra, 31006 Pamplona, Spain; andoni@unavarra.es (A.G.); sofia.korili@unavarra.es (S.K.)

* Correspondence: mavicente@usal.es (M.A.V.); emerson.faria@unifran.edu.br (E.H.d.F.); Tel.: +55-16-3711-8969 (E.H.d.F.)

Received: 19 March 2018; Accepted: 9 April 2018; Published: 11 April 2018

Featured Application: Zeolites obtained from natural clays by simple methods are potential efficient adsorbents of cationic dyes. Zeolites obtained from natural clays by simple methods may be used as efficient adsorbents of cationic dyes.

Abstract: The present work reports the synthesis of zeolites from two metakaolins, one derived from the white kaolin and the other derived from the red kaolin, found in a deposit in the city of São Simão (Brazil). The metakaolins were prepared by calcination of the kaolins at 600 °C; zeolite A was obtained after alkali treatment of the metakaolins with NaOH. The resulting solids were characterized by powder X-ray diffraction, thermal analysis, scanning electron microscopy, and nitrogen adsorption/desorption at −196 °C, which confirmed formation of zeolite A. The zeolites were applied as adsorbents to remove methylene blue, safranine, and malachite green from aqueous solutions. The zeolites displayed high adsorption capacity within short times (between one and five minutes); q_t was 0.96 mg/g. The equilibrium study showed that the zeolites had higher adsorption capacity for malachite green (q_e = 55.00 mg/g) than for the other two cationic dyes, and that the Langmuir isotherm was the model that best explained the adsorption mechanism.

Keywords: kaolin; metakaolin; zeolite A; cationic dye adsorption

1. Introduction

Industries such as dyes, textiles, paper, pharmaceuticals, and plastics generate a considerable amount of wastewater. Dyes are the primary contaminants recognized in wastewater due to their color. Approximately 15% of the total world production of dyes is lost during the dyeing process. Consequently, a large quantity of dyes is released into textile effluents [1]. Removing these contaminants from industrial wastewaters before effluents are discharged into the aquatic environment is obviously necessary. Various methods, such as biological processes [2], ultrasound [3], active carbon adsorption [4], ozone treatment [5], coagulation/flocculation [6], and ion exchange resin [7,8] and ion exchange membrane adsorption methods [9], have been investigated in an attempt to remediate dye-containing wastewaters. However, these methods are not always efficient and economical enough, so new alternative technologies that rely on low-cost materials are mandatory to solve the issues related to dye-containing wastewaters [10].

In this sense, natural clays have been modified to prepare several types of materials, including nanotubes [11], metakaolins [12], and zeolites [12]. The final material depends on the parent clay and on the treatment to which the clay is submitted. Zeolites obtained from kaolin are highlighted as advantageous, inexpensive materials for this purpose. Zeolites can be prepared by treating kaolin with alkaline compounds. The reactivity of the clay increases if it is calcined at medium temperatures (ca. 600 °C), to form metakaolins in the intermediate step [13].

Numerous researchers have revisited the presence of transition metal cations within the zeolite structure in recent years because these metal cations impart new properties to the solid, making them applicable for catalytic purposes or for environmental remediation through adsorption of various contaminants. For instance, Fe(III) cations, which very often exist as impurities in many zeolites, have been introduced into synthetic zeolites structurally or via cation exchange [14]. The presence of transition metal cations within the structure of a solid matrix can alter the crystallinity of the solid and even hinder clay conversion to zeolites in some cases [15].

Zeolites synthesized from several natural sources, such as clays and fly ash, are very important porous and selective adsorbents and display high ion exchange capacities. Table 1 lists some literature results on the preparation methods and on the conventional uses of zeolites.

Table 1. Literature data on the use of zeolites prepared from several sources as adsorbents of dyes from aqueous solutions.

Zeolite Type	Adsorbate	Adsorbed Amount (mg/g)	Reference
LTA (synthetic from fly ash)	Acid fuchsin	40.64	[16]
Zeolite (clinoptilolite-rich mineral)	Acid orange 95	3.4	[17]
Zeolite (clinoptilolite-rich mineral)	Acid blue 25	0.1	[18]
	Basic blue 9	82	
	Basic violet 3	98	
Zeolite (clinoptilolite mineral)	Methylene blue	82	[19]
	Methyl red	12	
Modified zeolite	Methylene blue	83	
	Methyl red	10	

Kaolin, which is very abundant in nature, is composed of kaolinite, a TO (tetrahedral–octahedral) clay mineral consisting of aluminum and silicon oxides; it also contains some iron, titanium, or manganese impurities as isomorphous substituents in both the octahedral and the tetrahedral layers or as extra framework phases, depending on where the clay originates. However, kaolin does not present good adsorption capacity due to its low cationic exchange capacity (CEC) and specific surface area (SSA) as compared to TOT (tetrahedral–octahedral–tetrahedral) clay minerals and other synthetic materials. Therefore, the synthesis of zeolites from clays such as kaolin is advantageous because zeolites have higher adsorption capacity [16,17]. In this context, we aimed to synthesize zeolites starting from two natural clay fractions as precursors, white kaolin and red kaolin (with higher iron content), and to investigate the potential of the resulting zeolites to adsorb the cationic dyes depicted in Table 2.

2. Experimental

2.1. Red and White Kaolin Purification

The parent clay minerals consisted of red and white kaolins obtained from a deposit in the city of São Simão, state of São Paulo, in the southeast of Brazil. The kaolins were supplied by the mining company Darcy R.O. Silva & Cia (São Simão-SP, Brazil). The natural clays were purified by the dispersion–decantation method. The purified clays were composed of very pure kaolinites and were designated as Ka (derived from white kaolin) and Ka-R (derived from red kaolin).

Table 2. Structures and relevant properties of the cationic dyes used in this work: methylene blue (MB), safranine (SA), and malachite green (MG).

Molecule	MB	SA	MG
Molecular dimension (Å)	$4.22 \times 13.19 \times 5.27$	$5.11 \times 11.84 \times 10.99$	$14.41 \times 4.23 \times 12.07$
Molecular surface (Å^2)	70	130	174
C.I. name	Basic blue 9	Basic red 2	Basic green 4
C.I.	52,015	50,240	42,000
Class	Thiazin	Safranin	Triarylmethane
λ_{max} (nm)	661	530	614, 425

2.2. Synthesis of Metakaolins

The purified kaolins Ka and Kao-R were calcined at 600 °C for 12 h, in air. The resulting metakaolins were designated M-Ka and M-Ka-R, respectively.

2.3. Synthesis of Zeolite A

The zeolites were synthesized through the hydrothermal route, which is a multiphase reaction–crystallization process, as emphasized by several authors [12–15,20]. This process usually encompasses at least one liquid phase as well as both amorphous and crystalline solid phases. Linde zeolite type A (LTA structure) was prepared by alkali treatment of M-Ka and M-Ka-R with aqueous NaOH 5 mol·L^{-1} solution at a NaOH/metakaolin molar ratio of 8:1; $Al_2Si_2O_7$ was considered the formula of the metakaolins [21]. The mixture was maintained under magnetic stirring and heated at 80 °C for 24 h. The resulting zeolites were washed with distilled water several times and dried at 110 °C. The zeolites obtained from M-Ka and M-Ka-R were labeled Zeo and Zeo-R, respectively.

2.4. Characterization Techniques

Methylene blue (MB) adsorption was used to quantify the CEC and the total SSA of the zeolites. A MB solution was prepared by dissolving 1 g of MB in 200 mL of distilled water. In parallel, 50 mg of the oven-dried zeolite was dispersed in 10 mL of distilled water. Then, consecutive volumes of 0.5 mL of the MB solution were added to the zeolite dispersion. After addition of each 0.5-mL aliquot of the MB solution, the mixture was homogenized by magnetic stirring for 1 min. Then, a drop was removed from the dispersion and placed onto Fisher brand filter paper, until a permanent blue halo of unadsorbed MB appeared on the filter paper. The SSA was determined from the MB amount required to reach the end-point, according to the equation

$$CEC = \frac{[MB] \times V}{W} \times 100 \tag{1}$$

where CEC is the cation exchange capacity (mEq·100 g^{-1}), [MB] is the MB solution concentration (mEq·L^{-1}), V (L) is the volume of MB solution used in the analysis, and W (g) is the mass of zeolite used in the experiment.

The SSA was calculated according to the Hang and Brindley method; Equation (2) was applied [22]

$$SSA = F_{MB} \times CEC \tag{2}$$

where SSA is the specific surface area ($m^2 \cdot g^{-1}$), and F_{MB} is the normalized value of the MB molecule surface area in m^2.

Powder X-ray diffraction (PXRD) patterns of non-oriented powder samples were recorded in a Siemens D-500 diffractometer (Siemens España, Madrid, Spain) with Ni-filtered Cu Kα radiation, working at 40 kV and 30 mA, at a scanning speed of 2°/min. Element chemical analyses were carried out by inductively coupled plasma–atomic emission spectrometry (ICP-AES) at Activation Laboratories Ltd. (Ancaster, ON, Canada). Infrared (FT-IR) spectra were recorded in the 350–4000 cm^{-1} range in a Spectrum-One Spectrometer (Perkin-Elmer, Waltham, MA, USA). The KBr pellet technique was employed: about 1 mg of the sample and 300 mg of KBr were used to prepare the pellets. Thermogravimetric (TG) and differential thermal (DTA) analyses were conducted in a SDT Q600 Simultaneous DTA-TGA-DSC equipment (Thermal Analysis TA, New Castle, PA, USA), from 25 to 900 °C, under nitrogen atmosphere, at a heating rate of 20 °C/min. The BET specific surface areas were determined from the corresponding nitrogen adsorption data at −196 °C, measured in an ASAP 2020 analyzer (Micrometrics, Norcross, GA, USA); 0.2 g of the sample was used. The samples were previously degassed at 200 °C for 24 h at a pressure lower than 50 μmHg. Scanning electron microscopy (SEM) photographs of the materials were obtained in a digital scanning microscope DSM 960 from (Zeiss, Carl Zeiss Iberia—Division Microscopy, Madrid, Spain); the samples had been coated with a thin gold layer by evaporation.

2.5. Adsorption Experiments

Equilibrium adsorption experiments were carried out in glass vials by shaking a known amount of the adsorbent, typically 0.1 g, with 5.0 mL of each dye solution at a given concentration, typically in the 0.14–1 $mmol \cdot L^{-1}$ range. After shaking for 30 min, a time selected from previous kinetic experiments, the solid phase was separated via centrifugation at 3500 rpm. The dye concentration in the solution was determined by UV–vis spectroscopy in a Hewlett-Packard 8453 Diode Array spectrometer (Agilent Technologies Spain, Madrid), and the amount of adsorbed dye was calculated by means of Equation (3)

$$q_e = \frac{(C_0 - C_e) \times V}{m} \tag{3}$$

where q_e ($mmol \cdot g^{-1}$) is the amount of adsorbed dye; C_i and C_e ($mmol \cdot L^{-1}$) are the initial and equilibrium liquid-phase concentrations of the dye, respectively; V (L) is the volume of the solution; and m (g) is the amount of adsorbent used in the experiment.

NaOH was supplied by Labsynth Products Laboratories (Diadema, Brazil). The other reagents used in the preparation and characterization of the zeolites and in adsorption experiments were purchased from Sigma-Aldrich® (Madrid, Spain) and had the highest purity, so they were used without purification. The gases were obtained from Oxi-Franca (Franca, Brazil).

3. Results and Discussion

3.1. Preparation and Characterization of the Materials

Table 3 lists the chemical composition of the parent kaolins Ka and Ka-R, which have very similar composition. Ka-R has slightly higher amounts of Fe_2O_3 and MnO than Ka. Clays usually have the reddish/brownish/yellowish color pattern on the surface of aggregates, on pore surfaces, and in the upper soil horizons as a result of oxidized transition metals such as Fe^{3+} and Mn^{4+}. The grayish/bluish colors of clays appear within aggregates or in the deeper soil horizons and are due to Fe^{2+} and Mn^{2+} cations present in the layers. The Fe^{3+} and Mn^{2+}/Mn^{4+} contents may justify the differences between the colors of Ka and Ka-R and should influence their reactivity depending on whether Fe^{3+} and Mn^{2+}/Mn^{4+} are present as extra framework phases or if they are located in structural positions.

The X-ray powder diffraction patterns of the zeolites Zeo and Zeo-R synthesized from the metakaolins M-Ka and M-Ka-R (Figure 1) reveal the presence of a single and very pure phase. Comparison with the

JCPDS file 43-142 clearly shows that Zeo and Zeo-R are hydrated Linde Type Zeolite A (LTA). LTA has a three-dimensional pore structure where the pores are perpendicular to each other along the x, y, and z directions. The pore diameter is defined by a small ring (4.2 Å) of eight oxygen atoms, which gives a large cavity with free diameter of 11.4 Å. The unit cell is cubic (a: 24.64 Å).

Table 3. Chemical composition (wt %) of the parent purified kaolins (Ka and Ka-R) and of the obtained zeolites (Zeo and Zeo-R), expressed on a dry basis (metal oxides 100%).

Sample	SiO_2	Al_2O_3	Fe_2O_3	MnO	MgO	CaO	Na_2O	K_2O	TiO_2
Ka	54.24	42.61	1.51	0.005	0.15	0.04	0.02	0.41	1.01
Ka-R	54.06	42.60	1.59	0.010	0.18	0.13	0.02	0.41	1.00
Zeo	43.52	34.39	0.67	0.005	0.11	0.03	20.59	0.14	0.55
Zeo-R	44.35	33.82	0.96	0.006	0.11	0.02	19.82	0.26	0.66

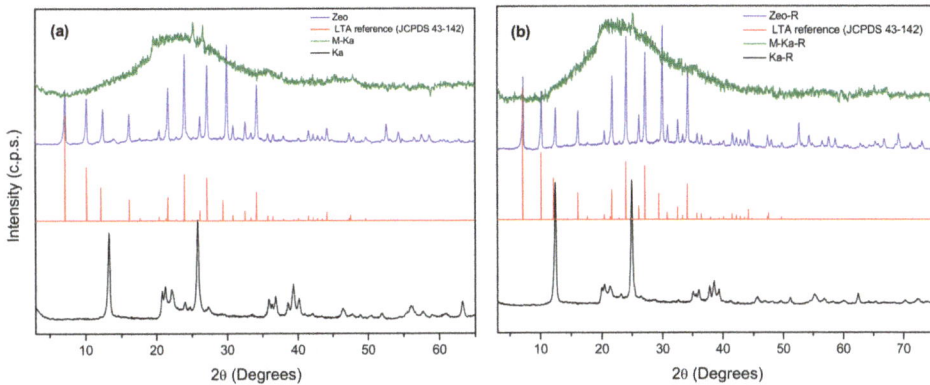

Figure 1. PXRD patterns of the purified parent kaolins Ka and Ka-R, the metakaolins M-Ka and M-Ka-R, and the synthesized zeolites Zeo and Zeo-R, prepared from white (**a**) and red (**b**) kaolins, respectively.

The purity of zeolites synthesized from kaolin has been reported to depend on various factors such as NaOH concentration, Al/Si ratio, calcination temperature of the parent kaolin, and duration of the treatment [12]. Here, the powder X-ray diffractograms prove that the small differences in Fe and Mn contents do not interfere in the synthesis of zeolite A: there is only a small decrease in the reflection intensities of the sample Zeo-R, in agreement with literature results. This decrease indicates the presence of isomorphic substituents in the clay mineral, which may lower the crystallinity of the resulting zeolite [15].

Thermogravimetric analyses (Figure 2) of the zeolites Zeo and Zeo-R give rise to more complex curves than the curves of the parent kaolins, Ka and Ka-R (not included, but typical of this mineral). Mass loss reaches values close to 23% for both Zeo and Zeo-R, which are very close to the expected value for zeolite A. The theoretical mass loss for a 'typical' zeolite A, with formula $Na_{12}Al_{12}Si_{12}O_{48} \cdot 27H_2O$, is 22.2%, but the mass loss strongly depends on the storage conditions. In the present case, special storage conditions were not adopted, so both zeolites Zeo and Zeo-R may have a low amount of adsorbed water. Water removal is almost complete around a temperature of 250 °C. The DTG curves (not shown) demonstrate that this removal occurs in three consecutive, overlapped processes centered at 65–70, 120–150, and 220–240 °C, which correspond to water held to the zeolite at various strengths, adsorbed water, and zeolitic water, respectively. Zeo-R undergoes a small additional mass loss at ca. 350 °C, which may be related to the removal of CO_2 fixed as carbonate during the preparation procedure in a strongly basic solution. At higher temperatures, dehydroxylation can be detected as a gentle, continuous mass decrease [23].

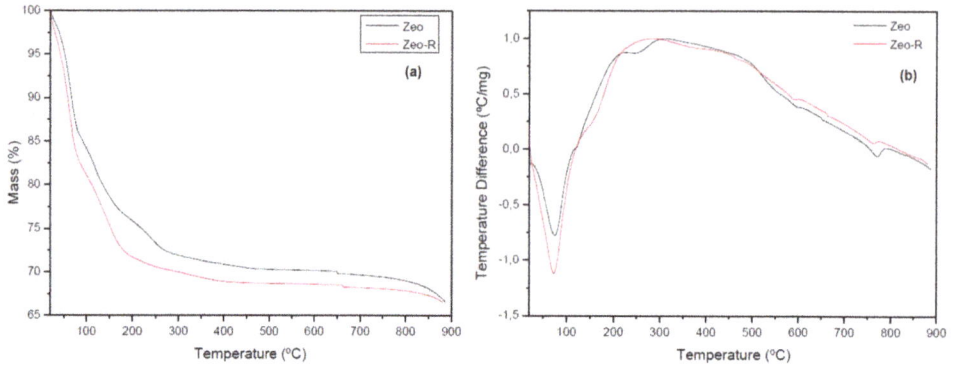

Figure 2. Thermogravimetric (**a**) and differential thermal (**b**) analysis curves of the zeolites prepared from white and red kaolins.

The nitrogen adsorption/desorption isotherms on the samples measured at −196 °C refer to type II isotherms (IUPAC classification [24]). This type of isotherm evidences that a complex structure with undefined pore size and volume size distributions exists due to low adsorption values (Figure 3). The BET SSA values follow the same trend in both series, increasing when one goes from kaolin to metakaolin and strongly decreasing when one goes from metakaolin to zeolite (Table 4). The increase from kaolin to metakaolin can be related to amorphization induced by calcination of kaolin to form metakaolin (see Figure 1). The sharp decrease from metakaolin to zeolite may be due to the extremely high crystallinity of the zeolite particles and their large particle size, as suggested by X-ray diffraction and confirmed by electron microscopy (vide infra). Given the large crystal size, the SSA is almost exclusively due to contribution from the external surface. In the same way, the pore size decreases when one goes from kaolin to metakaolin and increases when one goes from metakaolin to zeolite (Table 4).

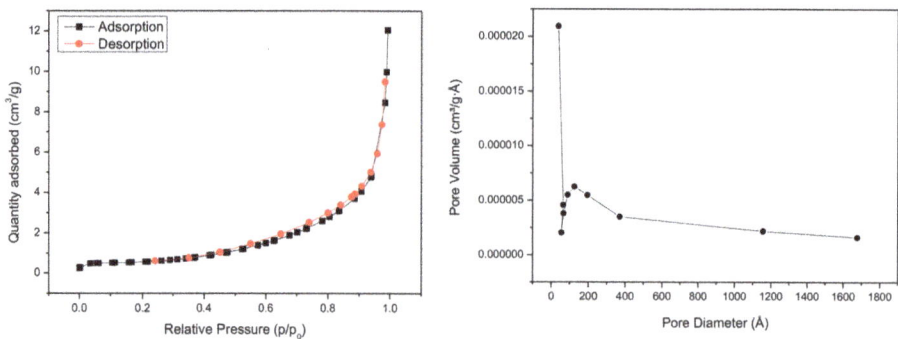

Figure 3. N_2 adsorption–desorption isotherm at −196 °C and BJH pore size distribution for the zeolite *Zeo*, derived from white kaolin.

Table 4. CEC and BET specific surface areas for the precursors Ka, Ka-R, M-Ka, and M-Ka-R; and the final zeolites Zeo and Zeo-R.

Sample	MB Saturation Volume (mL)	CEC (mEq·100 g^{-1})	CEC SSA [1] (m^2·g^{-1})	BET SSA [2] (m^2·g^{-1})	Pore Size (Å)
Ka	0.40	8.0	62	15	179
M-Ka	0.15	3.0	23	69	110
Zeo	0.40	8.0	62	2	355
Ka-R	0.30	6.0	47	16	184
M-Ka-R	0.10	2.0	16	79	113
Zeo-R	0.25	5.0	39	1	1130

[1] From CEC values obtained from MB adsorption; [2] From N_2 adsorption at $-196\ °C$.

Table 4 also provides the CEC SSA values regarding MB adsorption. In this case, CEC SSA decreases when one goes from kaolin to metakaolin and increases when one goes from metakaolin to zeolite. The CEC SSA values of the zeolites are very close to the CEC SSA values of the original kaolins. The CEC SSA values of Zeo and Zeo-R do not differ, which demonstrates that the slight differences in crystallinity do not affect CEC SSA. The results suggest that the CEC SSA of the kaolins Ka and Ka-R, which are originally rather low, decrease due to the strong calcination conditions applied to obtain the metakaolins, M-Ka and Me-Ka-R. Further reaction to produce the zeolites recovers the CEC SSA in the final Zeo and Zeo-R despite their large particle size and high crystallinity. However, the low surface area of the zeolites and the difficult access to their pores indicate that the experimental values may be affected by MB aggregation depending on the pH of the dispersions, to culminate in the large difference between the SSA values of the gas and liquid phase media.

The micrographs in Figure 4 depict the typical morphology of the very fine and rigid kaolinite particles. Stacked sheets forming tactoids, which produce aggregates, are evident. The micrographs of Zeo and Zeo-R reveal the formation of large cubic crystals, a typical morphology of these solids [25]. The shape of these crystals depends not only on the chemical composition of the zeolite, but also on the conditions used during the synthesis. In the current case, the morphology of the Zeo and Zeo-R crystals is very close to cubic. Zeo-R is more regular than Zeo. Aggregation of cubic crystal clusters gives quasi-spherical aggregates, an effect that is more evident for Zeo.

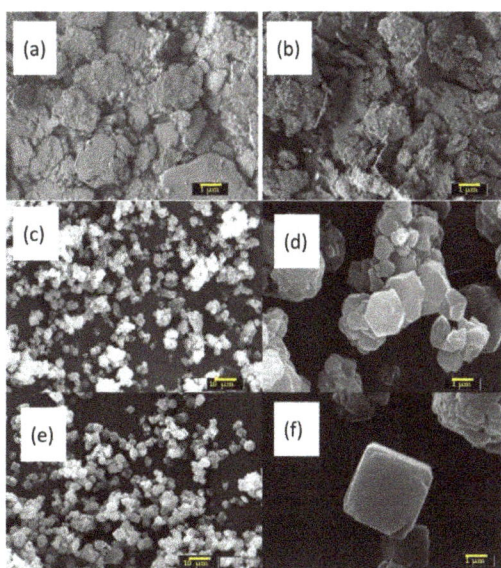

Figure 4. SEM micrographs of purified white kaolin (**a**) and red kaolin (**b**), Zeo zeolite (**c,d**) and Zeo-R zeolite (**e,f**).

3.2. Adsorption Studies

Equilibrium data, commonly denoted as adsorption isotherm, are important to define the mechanism of the adsorption process accurately [26]. Here, the Langmuir, Freundlich, and Sips isotherm models were used to analyze the adsorption of dyes onto Zeo and Zeo-R.

Figure 5 illustrates fitting of the experimental adsorption data with the aforementioned models. According to the Giles classification [27], the isotherms are type L-2 when MB and SA are tested as adsorbates, and type S-2 when MG is tested as adsorbate. L-type isotherms are characterized by a non-linear convex curve, which indicates lower availability of adsorption sites as the concentration at equilibrium increases and high affinity of the adsorbate for the adsorbent. As for S-type isotherms, the adsorption capacity of the solid increases as the adsorbate concentration rises.

Table 5 summarizes the parameters obtained from the nonlinear plots of each isotherm. Fitting of the models to the experimental data was evaluated by the chi-square test (χ^2) and the coefficient of determination (R). Most of the R values exceed 0.9 for the Langmuir model, which suggests that the model could fit the experimental results.

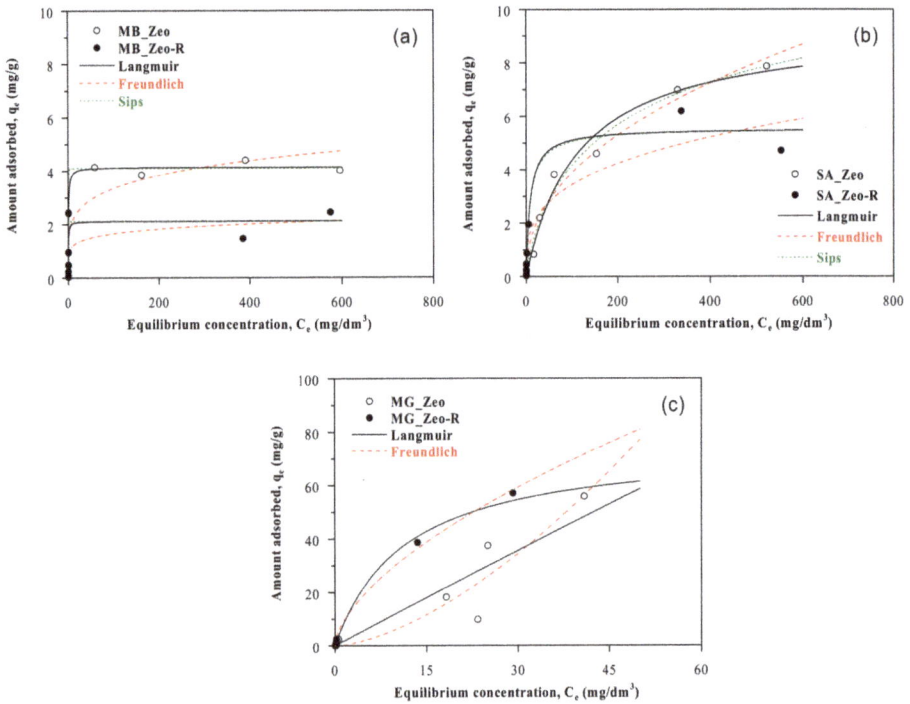

Figure 5. Equilibrium adsorption data for Zeo and Zeo-R as adsorbents and MB (**a**), SA (**b**), and MG (**c**) dyes as adsorbates.

The k_L values for the Langmuir model vary between 2.4 and 0.0006 L/mg. The k_L values are different when we compare the adsorption of dyes onto Zeo and Zeo-R (e.g., 2.4 and 1.4 L/mg for MB, respectively). Therefore, the adsorption capacity of the zeolites depends on the precise kaolin that is used as precursor for the synthesis of zeolites.

The decrease in the k_L values could be related to molecular volume and steric hindrance. This is not surprising because zeolite A has a narrow pore aperture, which makes access very difficult for the dye molecules [28]. The results indicate that the affinity of the binding sites in Zeo-R increases for all the dyes probably because there are isomorphic substitutions in the parent kaolin. The k_L values vary between zero and 0.11 for SA adsorption onto Zeo and Zeo-R, suggesting that this dye has large affinity for solids.

As previously reported by some authors, zeolite materials generally display high affinity for cationic dyes with very small molecular size, such as methylene blue. The behavior of MG, MB, and SA adsorption onto Zeo and Zeo-R shows that both zeolites have more affinity for MB, the dye with the smallest molecular size. Other authors have reported efficient removal of cationic dyes like basic blue 17, basic red 46, basic green 4, basic blue 9, basic violet 10, and basic yellow 28 by natural zeolites (e.g., clinoptilolite) [29–33]. In contrast, adsorption studies of anionic, acid dyes are very limited, and very low adsorption capacity values (0.5 mg g^{-1}) have been reported [18,34].

Table 5. Langmuir, Freundlich, and Sips equation parameters for the adsorption of MB, SA, and MG onto Zeo and Zeo-R.

Dye-Adsorbent	MB-Zeo	MB-Zeo-R	SA-Zeo	SA-Zeo-R	MG-Zeo	MG-Zeo-R
Langmuir [a]						
q_L (mg/g)	4.1	2.1	9.4	5.5	2094	75
k_L (L/mg)	1.4	2.4	0.0086	0.11	0.00057	0.087
χ^2	0.83	1.94	1.4	1.2	462	14
R	0.98	0.85	0.991	0.98	0.93	0.997
Freundlich [b]						
k_F (L/g)	1.4	0.86	0.49	0.85	0.16	7.4
m_F	5.2	7.1	2.2	3.3	0.63	1.6
χ^2	4.8	3.7	1.5	3.4	327	14
R	0.92	0.68	0.990	0.95	0.95	0.998
Sips [c]						
q_S (mg/g)	4.1	2.1	12	5.6	–	–
k_S (L/mg)	4.7	752	0.015	0.12	–	–
m_S	2.2	9.4	0.76	0.95	–	–
χ^2	0.34	0.91	1.2	1.2	–	–
R	0.995	0.93	0.992	0.98	–	–

[a] Langmuir equation isotherm: $q_e = \frac{q_L \cdot k_L \cdot C_e}{1 + k_L \cdot C_e}$; [b] Freundlich equation isotherm: $q_e = k_F \cdot C_e^{1/m_F}$; [c] Sips equation isotherm: $q_e = \frac{q_S \cdot k_S \cdot C_e^{m_S}}{1 + k_S \cdot C_e^{1/m_S}}$.

4. Conclusions

Zeolites obtained from kaolins efficiently adsorb methylene blue, safranine, and malachite green from aqueous solutions. The molecular volume of the dyes influences the adsorption capacity of zeolites and alters the adsorption rate calculated by theoretical kinetic and equilibrium models. In addition, the synthesized zeolites could have their adsorption capacity limited by the presence of cations and water molecules in their pores, which could block the access of larger molecules such as dyes. Overall, taking green chemistry requirements into account, zeolites prepared from natural clays are promising materials for the removal of many environmental pollutants.

Acknowledgments: The Spanish authors thank the Spanish Ministry of Economy and Competitiveness (MINECO) and the European Regional Development Fund (ERDF) for joint financial support (grant MAT2013–47811–C2–R and MAT2016-78863-C2-R). The Brazilian group acknowledges the support from the Brazilian research funding agencies Fundação de Amparo à Pesquisa do Estado de São Paulo (FAPESP) (311767/2015–0, 2012/08618-0, 2013/19523-3, and 2016/01501–1), Coordenação de Aperfeiçoamento de Pessoal de Nível Superior (CAPES), and Conselho Nacional de Desenvolvimento Científico e Tecnológico (CNPq). Both groups thank the Spain–Brazil Interuniversity Cooperation Grant, jointly funded by the Spanish Ministry of Education, Science and Sports (PHBP14/00003) and CAPES (317/15).

Author Contributions: E.J. Nassar, K.J. Ciuffi, M.A. Vicente, R. Trujillano, V. Rives, A. Gil, S. Korili and E.H. de Faria conceived, designed and guided the experiments; P.M. Pereira, B.F. Ferreira and N.P. Oliveira prepared the materials; all the authors participated in the characterization of the materials; P.M. Pereira, B.F. Ferreira and N.P. Oliveira wrote the first version of the manuscript; E.J. Nassar, K.J. Ciuffi, M.A. Vicente, R. Trujillano, V. Rives, A. Gil, S. Korili and E.H. de Faria revised/discussed the manuscript.

Conflicts of Interest: The authors declare no conflict of interest.

References

1. Fatima, M.; Farooq, R.; Lindström, R.W.; Saeed, M. A review on biocatalytic decomposition of azo dyes and electrons recovery. *J. Mol. Liq.* **2017**, *246*, 275–281. [CrossRef]
2. Kumar, S.S.; Shantkriti, S.; Muruganandham, T.; Murugesh, E.; Rane, N.; Govindwar, S.P. Bioinformatics aided microbial approach for bioremediation of wastewater containing textile dyes. *Ecol. Inform.* **2016**, *31*, 112–121.
3. Jawale, R.H.; Tandale, A.; Gogate, P.R. Novel approaches based on ultrasound for treatment of wastewater containing potassium ferrocyanide. *Ultrason. Sonochem.* **2017**, *38*, 402–409. [CrossRef] [PubMed]
4. Wong, S.; Yac'cob, N.A.N.; Ngadi, N.; Hassan, O.; Inuwa, I.M. From Pollutant to Solution of Wastewater Pollution: Synthesis of Activated Carbon from Textile Sludge for Dyes Adsorption. *Chin. J. Chem. Eng.* **2017**, in press. [CrossRef]
5. Shen, T.; Wang, Q.; Tong, S. Solid Base MgO/Ceramic Honeycomb Catalytic Ozonation of Acetic Acid in Water. *Ind. Eng. Chem. Res.* **2017**, *56*, 10965–10971. [CrossRef]
6. Liang, C.Z.; Sun, S.P.; Li, F.Y.; Ong, Y.K.; Chung, T.S. Treatment of highly concentrated wastewater containing multiple synthetic dyes by a combined process of coagulation/flocculation and nanofiltration. *J. Membr. Sci.* **2014**, *469*, 306–315. [CrossRef]
7. Edebali, S.; Pehlivan, E. Evaluation of chelate and cation exchange resins to remove copper ions. *Powder Technol.* **2016**, *301*, 520–525. [CrossRef]
8. Tan, L.; Shuang, C.; Wang, Y.; Wang, J.; Su, Y.; Li, A. Effect of pore structure on the removal of clofibric acid by magnetic anion exchange resin. *Chemosphere* **2018**, *191*, 817–824. [CrossRef] [PubMed]
9. Chitpong, N.; Husson, S.M. High-capacity, nanofiber-based ion-exchange membranes for the selective recovery of heavy metals from impaired waters. *Sep. Purif. Technol.* **2017**, *179*, 94–103. [CrossRef]
10. Huang, Z.; Li, Y.; Chen, W.; Shi, J.; Zhang, N.; Wang, X.; Li, Z.; Gao, L.; Zhang, Y. Modified bentonite adsorption of organic pollutants of dye wastewater. *Mater. Chem. Phys.* **2017**, *202*, 266–276. [CrossRef]
11. Dong, W.; Li, W.; Yu, K.; Krishna, K.; Song, L.; Wang, X.; Wang, Z.; Coppens, M.O.; Feng, S. Synthesis of silica nanotubes from kaolin clay. *Chem. Commun.* **2003**, 1302–1303. [CrossRef]
12. Belver, C.; Vicente, M.A. Easy Synthesis of K–F Zeolite from Kaolin, and Characterization of This Zeolite. *J. Chem. Educ.* **2006**, *83*, 1541–1542. [CrossRef]
13. Belver, C.; Bañares Muñoz, M.A.; Vicente, M.A. Chemical activation of a kaolinite under acid and alkaline conditions. *Chem. Mater.* **2002**, *14*, 2033–2043. [CrossRef]
14. Selim, M.M.; El-Mekkawi, D.M.; Aboelenin, R.M.M.; Sayed Ahmed, S.A.; Mohamed, G.M. Preparation and characterization of Na-A zeolite from aluminum scrub and commercial sodium silicate for the removal of Cd^{2+} from water. *J. Assoc. Arab Univ. Basic Appl. Sci.* **2017**, *24*, 19–25. [CrossRef]
15. Basaldella, E.I.; Sánchez, R.M.T.; Tara, J.C. Iron influence in the aluminosilicate zeolites synthesis. *Clays Clay Miner.* **1998**, *46*, 481–486. [CrossRef]
16. Xu, H.Y.; Wu, L.C.; Shi, T.N.; Liu, W.C.; Qi, S.Y. Adsorption of acid fuchsin onto LTA-type zeolite derived from fly ash. *Sci. China Technol. Sci.* **2014**, *57*, 1127–1134. [CrossRef]
17. Tümsek, F.; Avcı, Ö. Investigation of Kinetics and Isotherm Models for the Acid Orange 95 Adsorption from Aqueous Solution onto Natural Minerals. *J. Chem. Eng. Data* **2013**, *58*, 551–559. [CrossRef]
18. Hernández-Montoya, V.; Pérez-Cruz, M.A.; Mendoza-Castillo, D.I.; Moreno-Virgen, M.R.; Bonilla-Petriciolet, A. Competitive adsorption of dyes and heavy metals on zeolitic structures. *J. Environ. Manag.* **2013**, *116*, 213–221. [CrossRef] [PubMed]
19. Loannou, Z.; Karasawidis, C.; Dimirkou, A.; Antoniadis, V. Adsorption of methylene blue and methyl red dyes from aqueous solutions onto modified zeolites. *Water Sci. Technol.* **2013**, *67*, 1129–1136.
20. Nassar, M.Y.; Abdelrahman, E.A. Hydrothermal tuning of the morphology and crystallite size of zeolite nanostructures for simultaneous adsorption and photocatalytic degradation of methylene blue dye. *J. Mol. Liq.* **2017**, *242*, 364–374. [CrossRef]

21. El-Mekkawi, D.M.; Ibrahim, F.A.; Selim, M.M. Removal of methylene blue from water using zeolites prepared from Egyptian kaolins collected from different sources. *J. Environ. Chem. Eng.* **2016**, *4*, 1417–1422. [CrossRef]

22. Maček, M.; Mauko, A.; Mladenovič, A.; Majes, B.; Petkovšek, A. A comparison of methods used to characterize the soil specific surface area of clays. *Appl. Clay Sci.* **2013**, *83–84*, 144–152. [CrossRef]

23. Musyoka, N.M.; Petrik, L.F.; Hums, E.; Kuhnt, A.; Schwieger, W. Thermal stability studies of zeolites A and X synthesized from South African coal fly ash. *Res. Chem. Intermed.* **2015**, *41*, 575–582. [CrossRef]

24. Sing, K.S.W. Reporting physisorption data for gas/solid systems with special reference to the determination of surface area and porosity (Recommendations 1984). *Pure Appl. Chem.* **1985**, *57*, 603–619. [CrossRef]

25. Zide, D.; Fatoki, O.; Oputu, O.; Opeolu, B.; Nelana, S.; Olatunji, O. Zeolite "adsorption" capacities in aqueous acidic media; The role of acid choice and quantification method on ciprofloxacin removal. *Microporous Microporous Mater.* **2018**, *255*, 226–241. [CrossRef]

26. Moreira, M.A.; Ciuffi, K.J.; Rives, V.; Vicente, M.A.; Trujillano, R.; Gil, A.; Korili, S.A.; de Faria, E.H. Effect of chemical modification of palygorskite and sepiolite by 3-aminopropyltriethoxisilane on adsorption of cationic and anionic dyes. *Appl. Clay Sci.* **2017**, *135*, 394–404. [CrossRef]

27. Giles, C.H.; Smith, D.; Huitson, A. A General Treatment and Classification of the Solute Adsorption Isotherm. I. Theoretical. *J. Colloid Interface Sci.* **1974**, *47*, 766–778. [CrossRef]

28. Awala, H.; Leite, E.; Saint-Marcel, L.; Clet, G.; Retoux, R.; Naydenova, I.; Mintova, S. Properties of methylene blue in the presence of zeolite nanoparticles. *New J. Chem.* **2016**, *40*, 4277–4284. [CrossRef]

29. Alpat, S.K.; Özbayrak, Ö.; Alpat, S.; Akçay, H. The adsorption kinetics and removal of cationic dye, Toluidine Blue O, from aqueous solution with Turkish zeolite. *J. Hazard. Mater.* **2008**, *151*, 213–220. [CrossRef] [PubMed]

30. Karadag, D.; Turan, M.; Akgul, E.; Tok, S.; Faki, A. Adsorption equilibrium and kinetics of reactive black 5 and reactive red 239 in aqueous solution onto surfactant-modified zeolite. *J. Chem. Eng. Data* **2007**, *52*, 1615–1620. [CrossRef]

31. Wang, S.; Ariyanto, E. Competitive adsorption of malachite green and Pb ions on natural zeolite. *J. Colloid Interface Sci.* **2007**, *314*, 25–31. [CrossRef] [PubMed]

32. Wang, S.; Zhu, Z.H. Characterisation and environmental application of an Australian natural zeolite for basic dye removal from aqueous solution. *J. Hazard. Mater.* **2006**, *136*, 946–952. [CrossRef] [PubMed]

33. Yener, J.; Kopac, T.; Dogu, G.; Dogu, T. Adsorption of Basic Yellow 28 from aqueous solutions with clinoptilolite and amberlite. *J. Colloid Interface Sci.* **2006**, *294*, 255–264. [CrossRef] [PubMed]

34. Jin, X.; Jiang, M.Q.; Shan, X.Q.; Pei, Z.G.; Chen, Z. Adsorption of methylene blue and orange II onto unmodified and surfactant-modified zeolite. *J. Colloid Interface Sci.* **2008**, *328*, 243–247. [CrossRef] [PubMed]

© 2018 by the authors. Licensee MDPI, Basel, Switzerland. This article is an open access article distributed under the terms and conditions of the Creative Commons Attribution (CC BY) license (http://creativecommons.org/licenses/by/4.0/).

![applied sciences logo] *applied sciences*

MDPI

Article

Effect of Degassing on the Stability and Reversibility of Glycerol/ZSM-5 Zeolite System

Yafei Zhang [1,2], Rui Luo [3], Qulan Zhou [2,*], Xi Chen [4,5,*] and Yihua Dou [1]

[1] School of Mechanical Engineering, Xi'an Shiyou University, Xi'an 710065, China; effyzhang@126.com (Y.Z.); yhdou@vip.sina.com (Y.D.)
[2] State Key Laboratory of Multiphase Flow in Power Engineering, Xi'an Jiaotong University, Xi'an 710049, China
[3] Xi'an Thermal Power Research Institute Co., Ltd., Xi'an 710054, China; luorui@tpri.com.cn
[4] Department of Earth and Environmental Engineering, Columbia University, New York, NY 10027, USA
[5] International Center for Applied Mechanics, SV Lab, School of Aerospace, Xi'an Jiaotong University, Xi'an 710049, China
* Correspondence: qlzhou@mail.xjtu.edu.cn (Q.Z.); xichen@columbia.edu (X.C.); Tel.: +86-135-7199-5532 (Q.Z.); +86-187-1094-7000 (X.C.)

Received: 6 June 2018; Accepted: 26 June 2018; Published: 29 June 2018

Featured Application: A system composed of liquid and hydrophobic nanoporous materials can perform high energy absorption efficiency when applying external pressure. Such an advanced energy absorption system has the potential application for high-performance protection devices to protect the personnel and civil infrastructures from impact, and other areas that involve energy absorption/conversion. Gaseous phase in a liquid/nanoporous system plays roles during application. Understanding the effect of the amount of residual gas on the liquid/nanoporous system will help to guide the pretreatment of the liquid/nanoporous material mixture before encapsulating.

Abstract: Gaseous phase plays roles in a liquid/nanoporous system during application that adequate attention should be paid to the gaseous effects and the nanoscale gas-liquid interaction. In the present study, two glycerol/ZSM-5 zeolite systems with different amount of residual gas are compared by performing a series of experiments. Influences of loading rate, as well as system temperature on the gas-liquid interactions, are studied. Results show that vacuum degassing pretreatment is required to obtain a reversible and stable energy absorption system. Moreover, the influence of gas on a liquid/nanoporous system is found to mainly act on the liquid outflow. After the routine vacuum degassing pretreatment, the residual air that is left in the nanopores is around 0.9014 nm^{-3} per unit specific pore volume, as presented in the current study. During compression, the existing gas left in the nanochannel tends to gather into the gas cluster, which further promotes the liquid outflow during unloading. However, excessively dissolved gas may reduce the driving force for liquid outflow by breaking the continuity of the liquid molecular chain in nanochannel. Consequently, small bubbles as a labile factor in the system must be excluded for the steady use of the system. This work sheds some light on the effect of the amount of residual gas on the liquid/nanoporous system and gives guidance on the pretreatment of the liquid/nanoporous material mixture before encapsulating.

Keywords: liquid/nanoporous material system; gas amount; degassing pretreatment; liquid outflow; liquid-gas interaction

---

1. Introduction

When nonwetting liquid is forced into hydrophobic nanopores by applying external pressure, a large amount of mechanical energy can be converted to solid-liquid interfacial energy [1–6]. Due to the ultra-large solid-liquid interface, the pressure-induced intrusion of liquid into nanopores can contribute to a highly efficient energy absorption system [7,8]. Such advanced energy absorption system has the potential application for high-performance protection devices to protect the personnel and civil infrastructures from impact [9–13].

In comparison with the various types of materials developed for impact protection, the nonwetting liquid-hydrophobic nan-oporous system possesses the advantages of high energy absorption efficiency [14], ultra-fast energy dissipation rate [14], and fully reusability after up to thousands of cyclic loadings. However, not all of the tested nonwetting liquid-hydrophobic nanopores systems are reusable [15]. For a number of porous materials, it has been observed that the confined liquid tends to be "locked" inside the nanopores after intrusion [16,17], and may flow out again after certain irritation, such as lifting the system temperature [18].

For a steady use, the stability of the nonwetting liquid dispersed in the porous material has been studied in a number of works. Q Yu et al. [19] reported a jump phenomenon in a water and hydrophobic zeolites system and introduced a jump avoidance criterion through the frequency response function analysis. VD Borman et al. [20] studied the physical mechanism to describe the formation process of a stable state of nonwetting liquid filling porous material. Watt-Smith MJ et al. [21] constructed models representing the pore structures of amorphous, mesoporous silica pellets, and explained the physical processes of mercury retraction and entrapment in silica materials. Results showed that the portion of liquid confined in porous materials varies with the porosity, the average radius, the width of the pore size distribution, the concentration of substances in water, and the temperature.

To improve the system's reusability, experiments on the behaviors of nanoporous energy absorption systems that were subjected to cyclic loadings and the molecular dynamic simulations on mechanisms of liquid intrusion into nanopores have also been carried out. Adding chemical additives into water/nanopores systems can modify the inner surface of the porous material, change liquid-solid interaction, and improve the system's recoverability by assisting the liquid outflow. Kong et al. [22] found the system recoverability to be strongly dependent on the NaCl concentration. As NaCl concentration increased from 0% to 25.9 wt. %, the recoverability was improved by a factor of 3. Sun et al. [23] reported a defiltration control method of the zeolite ZSM-5/liquid system by adding sodium hydroxide (NaOH). They believed the infiltration of NaOH aqueous solutions would increase the density of silanol (Si-OH) groups on the inner surface of the porous ZSM-5, which helped to convert the originally hydrophobic surface to hydrophilic that ceased the liquid outflow. Lifting the system temperature is beneficial in establishing a recoverable and reusable energy absorption system. Correspondingly, the weakened intermolecular forces and temperature sensitivity of liquid viscosity may contribute to the infiltration and outflow process at a higher temperature level. Kong and Qiao [24] improved the system recoverability significantly by lifting system temperature in a water/hydrophobic mesoporous silica system. Zhang et al. [18] conducted cyclic loading experiment on a glycerol/ZSM-5 zeolite system, indicating that higher system temperature enable to enlarge the entry area of the nanochannels and to trigger the outflow of glycerol molecules. As gas is inevitable in a liquid/nanoporous system, gaseous phase trapped in nanoholes plays an important role in the system's reversibility. In 2007, Qiao et al. [25] reported that the gas-liquid interaction to be an indispensable factor in nanoenvironments. Through a molecular dynamics simulation, they found the gas molecules in relatively large nanochannels would be dissolved in the liquid during pressure-induced infiltration, leading to a so-called phenomenon of "nonoutflow", while gas molecules tended to form clusters in relatively small nanochannels by triggering liquid defiltration at a reduced pressure. Until 2014, Sun et al. [26] reported a series of experimental results on a water/zeolite β system, showing that the gas molecules acted as a dominant factor in affecting the liquid motion, and the gas nanophase's effect on water "outflow" was significantly time-dependent. Nano-confined liquid behavior exhibited unusual characteristics, such as the exceptionally high transport rate when compared with bulk fluid, which could not be explained by conventional continuum theories [27–31].

As gas plays roles in a liquid/nanoporous system during application, adequate attention should be paid to the gas phase effects and the nanoscale gas-liquid interaction. However, to the best of the authors' knowledge, the available researches on this topic are rather limited and the mechanisms of gas-liquid interaction are still unclear.

In this study, a series of comparative experiments are carried out on a degassed glycerol/ZSM-5 zeolite system and a normal glycerol/ZSM-5 zeolite system without degassing pretreatment. Keep in mind that the degassing treatment can only reduce the amount of residual gas in nanopores. In addition, the complete eliminating of gas in nanopores is almost impossible, and hence, two systems with different amount of residual gas are compared. Influences of the loading rate as well as the system temperature on the gas-liquid-solid interactions are studied. Results show that the degassing treatment is necessary to obtain a more reversible and stable energy absorption system.

2. Materials and Methods

2.1. Material Characterization

The ZSM-5 zeolite material was acquired from Shanghai Fuxu Molecular Sieve Limited Company in China. The particle size of the raw material is around 3.5 μm and the Na_2O concentration is 0.046%. To improve the purity of the sample material, the raw material is pretreated in a tube furnace under 873 K for six h. After decontamination pretreating, the properties of the zeolite are characterized by using Quantachrome Autosorb-iQ-MP gas sorption analyzer (Quantachrome Instruments, Boynton Beach, FL, USA). Its specific surface area and pore volume are determined to be 638.01 m^2/g and 510 mm^3/g, respectively. Through a Non-Local Density Functional Theory (DFT) analysis, both micro and meso pores are discovered in the sample. The micropore size is 0.524 nm and the mesopore size is around 2.114 nm. It is believed that the glycerol molecules only intrude into the mesopores (2.114 nm) under the infiltration pressure of 15~20 MPa in the present work, because the infiltration pressure of water molecules into MFI zeolite's micro pores around 0.5 nm are above 100 MPa, according to previous studies [5,32], and the size of glycerol molecule (about 0.62 nm) is larger than the size of water molecule. Although the glycerol molecules cannot intrude in the micropores (0.524 nm), the existence of micropores is of great importance for both infiltration and defiltration processes, especially for gaseous phase that is trapped in micropores, which cannot be neglected. So, the most probable pore size of mesopore (2.114 nm), the total surface area (638.01 m^2/g), and pore volume (510 mm^3/g) should be used for the calculation of relative parameters.

2.2. Sample Preparation

1.0 g of pretreated zeolite is suspended in 10 mL pure glycerol, forming a milklike liquid. Two conditions of glycerol/ZSM-5 zeolite systems are prepared with different amount of initial residual gas in the system. One condition, called the undegassed condition, is obtained by immersing zeolite into glycerol, and the mixture is placed in a dessicator for 12 h. The other condition, which is called degassed condition, is obtained by placing the glycerol/zeolite mixture in vacuum (0.003 MPa) for 12 h, so as to remove excessive air bubbles. Note that the degassing pretreatment could not eliminate gaseous phase completely, while there are still residual gas in nanopores. Degassing operation only changes the amount of gas that is left in nanopores when compared to the undegassed condition.

The after pretreated mixture is further sealed in a testing chamber, as illustrated in Figure 1. The volume of the chamber is 10 mL. The pressure could be applied by the piston with a diameter of 14 mm. The pressure holding ability of the chamber under different system temperatures is verified by keeping the pressure at the maximum for 600 s, with negligible pressure drop. Reference compression test of pure glycerol is carried out prior to testing any specific mixture. The difference between the mixture and reference is used to obtain the sorption isotherm.

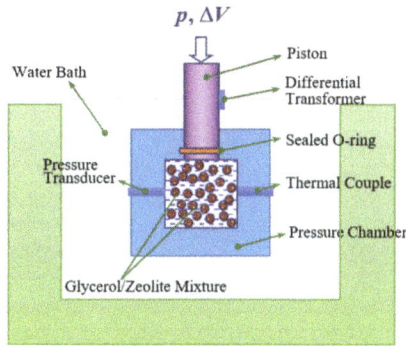

Figure 1. Schematic of experimental set-up.

2.3. Setup and Procedures of the Energy Absorption Experiment

A series of pressure-induced infiltration tests are conducted on a glycerol/ZSM-5 zeolite system. The schematic sketch of the experimental set-up is shown in Figure 1. The pressure-volume controlled testing chamber is a hexahedron cavity that is made of 1Cr13 stainless steel. In order to eliminate air bubbles during the sample filing procedure, the inner surfaces of the chamber are finely filleted. A polished steel piston is assembled in the center of the chamber's upper face, tightly sealed by an O-ring. The piston can be compressed into the chamber at various loading rates (0.01–0.1 mm/s) using a servo motor. Pressure in the chamber can be adjusted in the range of 0 to 60 MPa. The pressure chamber is immersed in a large electric-heated thermostatic water bath (from room temperature to 368 K) to gain control of the system temperature during test. A pressure transducer and a type-E thermal couple are attached to the front and back sides of the chamber to acquire the temperature and pressure of the system. A linear voltage differential transformer is attached on the piston to record its displacement. The real time pressure, temperature, and displacement signals are analyzed by an NI LabVIEW data acquisition system.

During experiment, the piston is moved downwards at a constant velocity in order to impose pressure on the mixture inside the chamber. Once the pressure reached the desired value, given as 50 MPa in this test, the piston starts to retract at the same rate. Under each working condition, the sample is compressed with ten successive loading-unloading cycles for a better observation on the system's stability and reversibility performances. P-ΔV curves (pressure-specific volume change curves) inside the pressure chamber are obtained to analyze the performance of the glycerol/zeolite system. P, which is defined as the pressure of the system, can be directly obtained from the pressure transducer. The specific volume change ΔV is defined as the volume change of the system per unit mass of zeolite and it equals the volume occupied by piston $A \times \Delta d$, where A and Δd is the cross-sectional area and displacement of the piston, respectively.

3. Results

The surface treated zeolite is lyophobic to glycerol, and thus an external pressure is required to induce the intrusion of glycerol molecules into the nanopores. When the external pressure retracts, the amount of glycerol molecules flowing out of the nanochanel is closely related to the state of gaseous phase in the system. In current study, two systems with different amount of residual gas are investigated, while the residual gas in nanopores after degassing pretreatment is estimated via test. Then, cyclic loading/unloading experiments are carried out on undegassed and degassed glycerol/zeolite systems under loading rate of 0.01 mm/s and 0.10 mm/s, with different system temperatures of 303 K and 348 K, respectively. Effects of the gas amount, the loading rate, and the system temperature are studied, and the mechanisms of gaseous phase on liquid outflow are also illuminated.

3.1. Residual Gas in Nanopores of Undegassed and Degassed Mixture

Figure 2 are photos of glycerol/zeolite mixture taken after different treatment conditions, which show the gas content of the system before and after degassing treatment. Figure 2a is taken after glycerol/zeolite mixture placing at room temperature in a drying condition for 12 h, which demonstrates the gas content of an undegassed system. It is seen that there are a great amount of air bubbles suspended in the mixture. These small bubbles cannot surface because of the viscosity resistance of the liquid. Figure 2b is taken after placing the glycerol/zeolite mixture in vacuum for 12 h, representing the gas content of a degassed system. After 12 h' vacuum degassing, there are no visible bubbles in the system. However, we still carry out a more strict residual air test to estimate the amount of air that stayed in the zeolite's nanopores for degassed mixtures. The following paragraphs are the detailed procedural and analysis of the residual air test of the degassed system. Figure 2c is taken after the residual air test. The bubbles on the surface of the mixture prove that the used degassing treatment under normal temperature cannot eliminate air in nanopores completely.

(a) undegassed mixture   (b) degassed mixture   (c) degassed mixture after heating in vacuum

Figure 2. Photos of glycerol/zeolite mixture taken after different treatment.

The residual air test is conducted with a DZF-6000 vacuum drying oven (Xi'an Yu Hui Experimental Apparatus, Xi'an, China). Put the degassed mixture in the vacuum drying oven, and vacuumize the oven to the maximum vacuum degree (0.003 MPa). Then, start the heating schedule of the oven to elevate the inside temperature. With the increase of temperature, the residual air in zeolite will get out of the porous channel gradually; meanwhile, the changing temperature and the pressure of the vacuum drying oven are recorded. Thus, the amount of air that is released from nanopores under certain temperatures can be calculated through the ideal gas law.

The real gas can be considered as ideal gas when both molecular size and inter molecular attractions could be neglected. The test temperature is from 303.15 to 373.15 K and the test pressure is from approximate vacuum to atmosphere. For a large volume at lower pressures, because the average distance between adjacent molecules becomes much larger than the molecular size, the neglect of molecular size becomes less important for lower densities. On the other side, with increasing temperatures, the relative importance of intermolecular attractions diminishes with increasing thermal kinetic energy. Therefore, the ideal gas law can be applied.

Since the absolute vacuum is unavailable, there is still air left in the oven at the highest vacuum degree. In addition, the vacuum drying oven may leak during heating process. A reference test with the same volume of pure glycerol is carried out. For both reference test and test with the degassed mixture, the temperature inside the oven is set to increase from 303.15 to 373.15 K, and the pressure variation is recorded during heating. Both pure glycerol and glycerol/zeolite mixture are vacuum degassing pretreated before test.

For reference test, there is:

$$pV_0 = n_0RT \qquad (1)$$

where, p is the absolute pressure inside the vacuum drying oven, V_0 is the volume of the vacuum drying oven, which is a constant and can be measured in advance, n_0 is the moles of air released during reference test, R is the ideal gas constant, and T is the temperature inside the vacuum drying oven.

For test with glycerol/zeolite mixture, there is:

$$pV_0 = (n_0 + \Delta n)RT \qquad (2)$$

where, Δn is the amount of air released from nanopores under a certain temperature.

The n-T curve of reference test and test with degassed glycerol/zeolite mixture is graphed in Figure 3.

From Figure 3, it is clearly observed that there is 4.642×10^{-4} mol·g^{-1} air released from nanopores under 373.15 K. As the specific pore volume for pristine zeolite is about 0.31 cm^3·g^{-1}, there are at least 0.9014 nm^{-3} per unit specific pore volume air still in zeolite after vacuum degassing pretreatment.

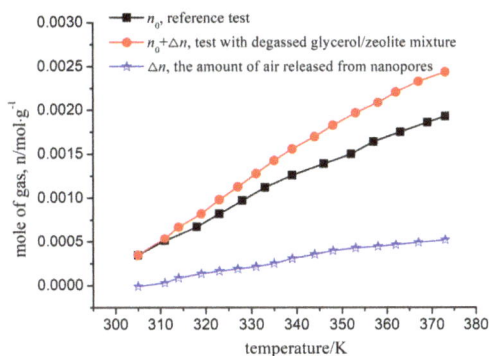

Figure 3. The amount of air released from nanopores with the increase of temperature after degassing pretreatment.

It is with regret that the above method is not suitable for estimating residual air in the undegassed system, since the amount of gas during vacuumizing at the very beginning cannot be quantified.

3.2. Effect of Gas Amount under Different Loading Rates and System Temperatures

3.2.1. P-ΔV Curves of Energy Absorption Experiment

Figure 4 shows the P-ΔV curves (pressure-specific volume change curves) of both undegassed and degassed glycerol/zeolite systems under static/fast loading rate and at the room/lifted system temperature. At the beginning of loading, glycerol stayed outside the hydrophobic nanopores at relative lower pressure. P changes linearly with ΔV in this stage. Once the pressure rises beyond the critical infiltration pressure, the glycerol molecules overcomes the capillary resistance and flows into the nanopores, forming an infiltration plateau. When the nanopores are filled, the system compressibility featured the linear compression behavior of liquid filled solid phase. During the unloading process, glycerol molecules flow out of the nanopores until the pressure is below a critical defiltration pressure. When the external pressure is reduced back to 0.1 MPa, the isotherm curve does not completely return to its origin. Different degrees of outflow of the invaded glycerol molecules perform under different working conditions.

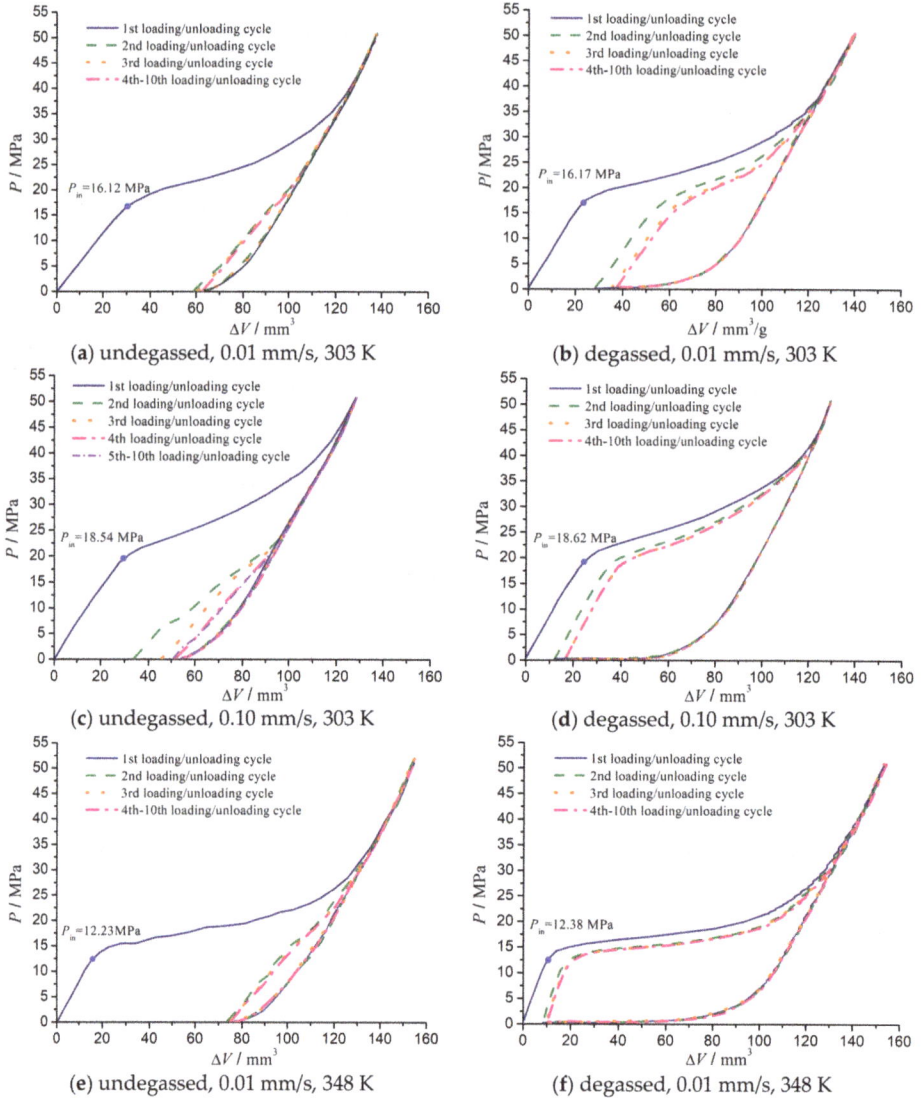

Figure 4. *P-ΔV* curves of undegassed and degassed glycerol/zeolite system under ten cyclic loadings (loading rate 0.01 mm/s and 0.10 mm/s; system temperature 303 K and 348 K).

3.2.2. Comparison of *P-ΔV* Curves in Comparative Working Condition

Figure 4a,c show the 10 cycles of *P-ΔV* curves of the undegassed glycerol/zeolite system at 0.01 mm/s and 0.10 mm/s, respectively. At both loading rates, the glycerol/zeolite system without degassing pretreatment acts as a disposable system. After the 1st intrusion, a large amount of accessible nanopore space is permanently occupied by the invaded glycerol molecules, thus little infiltration is allowed in subsequent cycles. In Figure 4c, although a higher loading rate leads a better outflow after the first intrusion, the hysteresis loop of the undegassed system shrinks afterwards.

Figure 4b and d show the 10 cycles of P-ΔV curves of the degassed glycerol/zeolite system at 0.01 mm/s and 0.10 mm/s separately. When comparing with the undegassed condition, eliminating excessive air bubbles during preparing glycerol/zeolite mixture promotes the reversibility of the system dramatically. In the degassed system, although part of the glycerol molecules trapped in the nanopores and could not flow out of the channels freely after the first cycle, the system reaches its thorough balance, and performs a pressure plateau after the first two or three cycles.

Furthermore, the P-ΔV curves of the degassed and the undegassed systems at a lifted system temperature of 348 K are obtained. The results are plotted in Figure 4e,f. As shown in Figure 4f, the stability of the degassed system is still good at higher system temperature, and the reversibility of the degassed system is improved as well. The system achieves its throughput balance after three cycles. While for an undegassed system in Figure 4e, the P-ΔV curves during loading processes fluctuate, indicating an unstable intrusion process. This is caused by the enhanced solubility of gaseous phase under higher system temperature. The detailed liquid-gas interaction will be analyzed in Section 4, the discussion section.

3.2.3. Effect of Gas Amount on Liquid Outflow

Extract the first loading/unloading cycle under various working conditions, and plot them together in Figure 5. For the first loading process, the accessible pore volume is the same for both the degassed and undegassed system. Although with different amount of residual gas in nanopores, there is no big difference for the first intrusion process at a certain contrast working condition because of the strong compressibility of gaseous phase. However, the disparate degrees of outflow phenomenon of systems with different gas amount prove that residual gas plays quite an important role in liquid outflow.

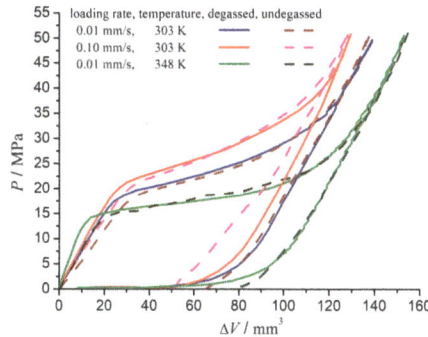

Figure 5. Comparision of the first loading/unloading cycle of undegassed and degassed glycerol/zeolite system (loading rate 0.01 mm/s and 0.10 mm/s; system temperature 303 K).

Define infiltration percentage as $\Delta V_{in}^{i}/\Delta V_{in}^{1}$ and defiltration percentage as $\Delta V_{de}^{i}/\Delta V_{in}^{1}$, where ΔV_{in}^{i} is the inflow volume of liquid during the ith intrusion process, and ΔV_{de}^{i} is the outflow volume of liquid after the ith loading/unloading cycle. During the first loading/unloading process, not the entire nanochannel will be filled with glycerol molecules, yet part of the glycerol molecules is trapped in the nanopores and could not flow out of the channels freely after the first cycle. Defects may be produced upon intrusion-extrusion cycles, by breaking Si-O-Si bonds into Si-OH (silanol) groups [21,33]. The incomplete outflow of the 1st cycle is partly due to trapped glycerol molecules and partly due to inelastic deformation of the system [18]. Assuming the infiltration volume of the first intrusion process ΔV_{in}^{1} as the maximum accessible infiltration volume, ΔV_{in}^{1} is taken as the baseline of measurement.

The infiltration and defiltration percentages of 10 loading/unloading cycles are portrayed in Figure 6. Both the infiltration and defiltration percentages of a degassed system are larger than that of an undegassed system under a certain contrast working condition. As the difference between the degassed and undegassed system mainly ascribes to the amount of residual gas inside nanopores, it can be concluded that certain amount of residual gas helps liquid to flow out of nanopores, in turn to elevate the system's reversibility. When comparing the infiltration and defiltration percentage, more glycerol molecules tend to flow in and out of the nanochannels under higher loading rate, and at a lifted system temperature.

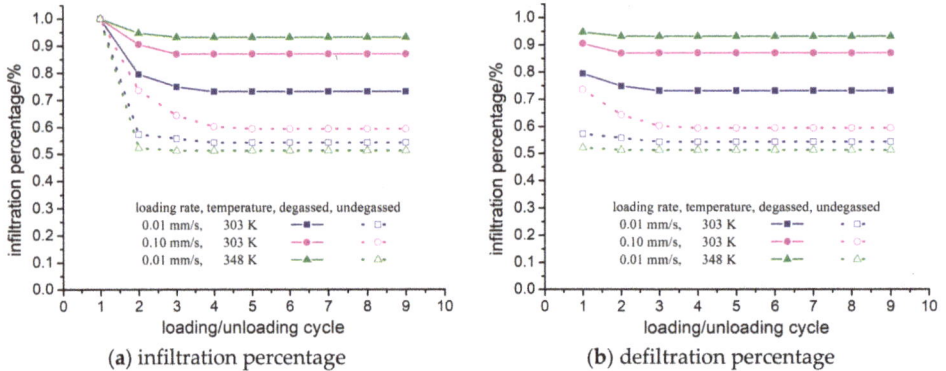

Figure 6. Comparison of the infiltration and defiltration percentage of undegassed and degassed glycerol/zeolite system (loading rate 0.01 mm/s and 0.10 mm/s; system temperature 303 K and 348 K).

4. Discussion

When considering the different amounts of residual gas inside the system, the above phenomenon is proof of understanding the gas behavior under finite circumstances. In general, the gas in a liquid/nanoporous system can be divided into three categories: small bubbles standing in the mixture, the gas dissolved in the liquid, and the gas trapped in pores. For the undegassed condition, although the mixture are stood in atmosphere for 12 h during pretreatment, small bubbles that were introduced in the process of mixing glycerol and zeolite powder would not surface up spontaneously, as well as the gas dissolved and the gas trapped in pores. Therefore, the three categories of gas mentioned above are all included in the undegassed system. For the degassed mixture, as has been placed in vacuum for 12 h, the small bubbles dispersing in the mixture, together with the gas dissolved in the liquid as well, could be excluded. As a consequence, only the gas residual in the nanopores could not be eliminated during degassing pretreatment. To note that after pretreatment and before sealing, when the system is taken out of the vacuum condition, air would dissolved in liquid under atmospheric pressure. Assuming that in both degassed and undegassed cases, the amount of dissolved gas in the liquid are the same and equal to the solubility of air under atmospheric pressure. Therefore, the comparative differences between the above two working conditions are very obvious: there are small bubbles in the undegassed system (referring to Figure 2a), while none in the degassed system (referring to Figure 2b).

During intrusion and extrusion, the analyzed three categories of residual gas play different roles, which mainly depend on where the gas is trapped and how easily it dissolves in the liquid phase. According to Qiao's study, gas molecules tend to form clusters in nanochannels less than a few nanometers, which triggers liquid defiltration at a reduced pressure [25]. Since a molecular-sized nanoscale confinement arranges the gas and the water molecules into a single-file chain inside molecular-sized nanochannels [27,34], the radical confinement therefore provides a high resistance for molecular site exchanges [26]. Thus, the residual gas in nanopores under either undegassed or degassed systems helps to trigger the glycerol outflow. For a system spreading with small bubbles, although the gas solubility would increase with the increasing pressure, nevertheless, the undissolved gaseous phase would inevitably be pressed into the nanochannel during loading process, to form a long or short section of gas cluster in the nanochannel. As the liquid is abundant in the testing systems, there is always liquid entering the channel. A cartoon depicting the status of gas-liquid in a nanochannel in the current system is given in Figure 7. Firstly, on the inside of the nanochannel, there is always a gas cluster consisting of the gas molecules that exists in nanopores originally. In a degassed system, a liquid column/chain fills the rest of the nanochannel until the open end, as depicted in Figure 7a. In an undegassed system, liquid column and gas cluster or gas molecule exist alternately with each other, as depicted in Figure 7b. In a hydrophobic system, the repelling force of the rigid wall to liquid is one of the most important driving forces for outflow during unloading. For instance, water in the hydrophobic zeolite behaves as a nanodroplet trying to close its hydrogen bonds onto itself, with a few short-lived dangling OH groups [35]. However, there is no such repelling force between the rigid wall and gas molecules. The inserted gas molecule interrupts the continuity of liquid chain, and the attractive intermolecular forces between liquid–liquid molecules is replaced by the repulsive force between the liquid and the gas molecules. Due to the lack of repelling force from the rigid wall and the insert of the repulsive force between liquid and gas molecules, the driving force for outflow decreases. Thus, the liquid phase tends to be trapped inside the nanochannel permanently. The trapped state of the nonwetting liquid appears as a result of a thermodynamic equilibrium among the interaction of the neighboring liquid-gas clusters and the liquid-solid interface.

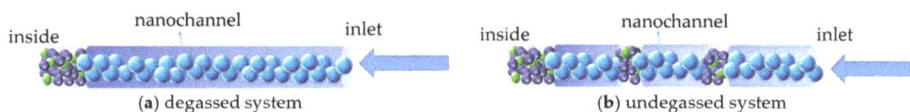

Figure 7. Cartoon of status of gas liquid in a nanochannel of degassed and undegassed systems. (🔵 gas cluster, 〰️ liquid colum/chain.

In the undegassed system, the alternative distribution of the liquid column/chain and gas cluster/molecule is anisotropic and energetically unfavorable or unstable. The system stability is especially poor for the undegassed system under a higher loading rate (0.1 mm/s) and system temperature (348 K), as shown in Figure 4c,e. The fluctuations of the P-ΔV curves are probably caused by the uncompleted dissolution of small bubbles with a fast increased pressure under a high loading rate. The alternative distribution of the liquid column and the gas cluster is more likely to be formed in a nanochannel rather than a liquid column with gas molecule inserted distribution. Besides, this distribution is more unfavorable and energetically unstable. Through molecular thermal motion, gas, and liquid molecules adjacent will keep colliding with each other with high velocity and frequency in a random way, and hence, a gradual gas-liquid molecular site exchange can occur, despite in a much slower way than the diffusion process in the bulk phase in the sense of probability statistics. The higher the system temperature, the more furiously molecules collide each other, and the less stable the system is. In relatively large pores, gas channeling may also occur. These could be factors causing the fluctuation of P-ΔV curves in Figure 4e when the system temperature is 348 K.

To sum up, the influence of gas on a liquid/nanoporous system is mainly on the liquid outflow. On one hand, the originally existing gas in the nanochannel is compressed into a gas cluster, which will promote the liquid outflow during unloading. On the other hand, the excessively dissolved gas may break the continuity of the liquid molecular chain in the nanochannel and then the driving force for liquid outflow is reduced correspondingly, which in turn reduces the system's reversibility. Moreover, the increase of the amount of gas in the system will lead to instability if the excess small bubbles are not dissolved in time, especially under working conditions of a high loading rate or high system temperature.

5. Conclusions

In this study, a series of comparative experiment are carried out on a degassed glycerol/ZSM-5 zeolite system and an undegassed glycerol/ZSM-5 zeolite system. The effects of degassing on the stability and reversibility of glycerol/ZSM-5 zeolite system are studied by comparing the P-ΔV characteristics of different systems by varying the amount of residual gas, the loading rate, and the system temperature. Results show that it is impossible and unnecessary to fully eliminate the residual air in the nanopores by routine vacuum degassing pretreatment. Adequate residual air in nanopores, which is around 0.9014 nm^{-3} per unit specific pore volume air in the current study, is beneficial to elevate the system's reversibility. However, the vacuum degassing treatment is necessary to obtain a more reversible and stable energy absorption system. The originally existing air in nanopores is necessary to help the liquid outflow, nevertheless, excessively dissolved gas may reduce the driving force on liquid outflow by breaking the continuity of the liquid molecular chain in the nanochannel, let alone small bubbles, which are labile factor for the system's steady use. This work sheds some light on the effect of the residual amount of gas on the liquid/nanoporous material system and it gives us some guidance on the pretreatment of the liquid/nanoporous material mixture before encapsulating.

Author Contributions: Conceptualization, Y.Z., Q.Z. and X.C.; Methodology, Y.Z. and R.L.; Validation, Y.Z.; Formal Analysis, Y.Z. and R.L.; Resources, Q.Z. and Y.D.; Writing—Original Draft Preparation, Y.Z.; Writing—Review & Editing, R.L. and Y.D.; Supervision, Q.Z. and X.C. Funding Acquisition, Q.Z., Y.Z. and Y.D.

Funding: This research was funded by [Natural Science Foundation of Shaanxi Province of China] grant number [2017JQ5108], [Special Scientific Research Plan of Shaanxi Province Education Department] grant number [17JK0594], [National Science and Technology Major Project of the Ministry of Science and Technology of China] grant number [2016ZX05017-006-HZ03], [Ministry of Industry and Information Technology Support Project for High-Tech Ships] and [the Fundamental Research Funds for the Central Universities in Xi'an Jiaotong University].

Conflicts of Interest: The authors declare no conflict of interest. The funders had no role in the design of the study; in the collection, analyses, or interpretation of data; in the writing of the manuscript, and in the decision to publish the results.

References

1. Eroshenko, V.; Regis, R.; Soulard, M.; Patarin, J. Energetics: A new field of applications for hydrophobic zeolites. *J. Am. Chem. Soc.* **2001**, *123*, 8129–8130. [CrossRef] [PubMed]
2. Chen, X.; Qiao, Y. Science and prospects of using nanoporous materials for energy absorption. In *Life-Cycle Analysis for New Energy Conversion and Storage Systems*; Fthenakis, V., Dillon, A., Savage, N., Eds.; Cambridge University Press: Cambridge, MA, USA, 2008; Volume 1041, pp. 95–105.
3. Zhao, J.; Qiao, Y.; Culligan, P.J.; Chen, X. Confined liquid flow in nanotube: A numerical study and implications for energy absorption. *J. Comput. Theor. Nanosci.* **2009**, *7*, 379–387. [CrossRef]
4. Cao, G. Working mechanism of nanoporous energy absorption system under high speed loading. *J. Phys. Chem. C* **2012**, *116*, 8278–8286. [CrossRef]
5. Sun, Y.; Xu, J.; Li, Y.B.; Liu, B.; Wang, Y.; Liu, C.; Chen, X. Experimental study on energy dissipation characteristics of ZSM-5 zeolite/water system. *Adv. Eng. Mater.* **2013**, *15*, 740–746. [CrossRef]
6. Khay, I.; Chaplais, G.; Nouali, H.; Ortiz, G.; Marichal, C.; Patarin, J. Assessment of the energetic performances of various zifs with sod or rho topology using high pressure water intrusion-extrusion experiments. *Dalton Trans.* **2016**, *45*, 4392–4400. [CrossRef] [PubMed]
7. Sun, Y.; Guo, Z.; Xu, J.; Xu, X.; Liu, C.; Li, Y. A candidate of mechanical energy mitigation system: Dynamic and quasi-static behaviors and mechanisms of zeolite β/water system. *Mater. Des.* **2015**, *66*, 545–551. [CrossRef]
8. Li, M.; Lu, W. Liquid marble: A novel liquid nanofoam structure for energy absorption. *AIP Adv.* **2017**, *7*, 055312. [CrossRef]
9. Chen, X.; Surani, F.B.; Kong, X.; Punyamurtula, V.K. Energy absorption performance of steel tubes enhanced by a nanoporous material functionalized liquid. *Appl. Phys. Lett.* **2006**, *89*, 241918. [CrossRef]
10. Liu, Y.; Schaedler, T.A.; Jacobsen, A.J.; Lu, W.; Qiao, Y.; Chen, X. Quasi-static crush behavior of hollow microtruss filled with nmf liquid. *Compos. Struct.* **2014**, *115*, 29–40. [CrossRef]

11. Sun, Y.; Li, Y.; Zhao, C.; Wang, M.; Lu, W.; Qiao, Y. Crushing of circular steel tubes filled with nanoporous-materials-functionalized liquid. *Int. J. Damage Mech.* **2018**, *27*, 439–450. [CrossRef]

12. Sun, Y.; Xu, X.; Xu, C.; Qiao, Y.; Li, Y. Elastomeric cellular structure enhanced by compressible liquid filler. *Sci. Rep.* **2016**, *6*, 26694. [CrossRef] [PubMed]

13. Xu, J.; Hu, R.; Chen, X.; Hu, D. Impact protection behavior of a mordenite zeolite system. *Eur. Phys. J. Spec. Top.* **2016**, *225*, 363–373. [CrossRef]

14. Li, M.; Lu, W. Adaptive liquid flow behavior in 3d nanopores. *Phys. Chem. Chem. Phys.* **2017**, *19*, 17167–17172. [CrossRef] [PubMed]

15. Surani, F.B.; Kong, X.; Panchal, D.B.; Qiao, Y. Energy absorption of a nanoporous system subjected to dynamic loadings. *Appl. Phys. Lett.* **2005**, *87*, 163111. [CrossRef]

16. Han, A.; Punyamurtula, V.K.; Qiao, Y. Heat generation associated with pressure-induced infiltration in a nanoporous silica gel. *J. Mater. Res.* **2008**, *23*, 1902–1906. [CrossRef]

17. Lefevre, B.; Saugey, A.; Barrat, J.L.; Bocquet, L.; Charlaix, E.; Gobin, P.F.; Vigier, G. Intrusion and extrusion of water in hydrophobic mesopores. *J. Chem. Phys.* **2004**, *120*, 4927. [CrossRef] [PubMed]

18. Zhang, Y.; Li, N.; Luo, R.; Zhang, Y.; Zhou, Q.; Chen, X. Experimental study on thermal effect on infiltration mechanisms of glycerol into ZSM-5 zeolite under cyclic loadings. *J. Phys. D Appl. Phys.* **2016**, *49*, 025303. [CrossRef]

19. Yu, M.C.; Gao, X.; Chen, Q. Nonlinear frequency response analysis and jump avoidance design of molecular spring isolator. *J. Mech.* **2016**, *32*, 527–538. [CrossRef]

20. Borman, V.D.; Belogorlov, A.A.; Byrkin, V.A.; Tronin, V.N.; Troyan, V.I. Stability of a nonwetting liquid in a nanoporous medium. *Phys. Scr.* **2014**, *89*. [CrossRef]

21. Watt-Smith, M.J.; Rigby, S.P.; Chudek, J.A.; Fletcher, R. Simulation of nonwetting phase entrapment within porous media using magnetic resonance imaging. *Langmuir* **2006**, *22*, 5180–5188. [CrossRef] [PubMed]

22. Kong, X.; Qiao, Y. Improvement of recoverability of a nanoporous energy absorption system by using chemical admixture. *Appl. Phys. Lett.* **2005**, *86*, 151919. [CrossRef]

23. Sun, Y.; Lu, W.; Li, Y. A defiltration control method of pressurized liquid in zeolite ZSM-5 by silanol introduction. *Appl. Phys. Lett.* **2014**, *105*, 121609. [CrossRef]

24. Kong, X.; Qiao, Y. Thermal effect on pressure induced infiltration of a nanoporous system. *Phil. Mag. Lett.* **2005**, *85*, 331–337. [CrossRef]

25. Qiao, Y.; Cao, G.; Chen, X. Effects of gas molecules on nanofluidic behaviors. *J. Am. Chem. Soc.* **2007**, *129*, 2355–2359. [CrossRef] [PubMed]

26. Sun, Y.; Li, P.; Yu, Q.; Li, Y. Time-dependent gas-liquid interaction in molecular-sized nanopores. *Sci. Rep.* **2014**, *4*, 6547. [CrossRef] [PubMed]

27. Hummer, G.; Rasaiah, J.C.; Noworyta, J.P. Water conduction through the hydrophobic channel of a carbon nanotube. *Nature* **2001**, *414*, 188–190. [CrossRef] [PubMed]

28. Majumder, M.; Chopra, N.; Andrews, R.; Hinds, B.J. Nanoscale hydrodynamics: Enhanced flow in carbon nanotubes. *Nature* **2005**, *438*, 44. [CrossRef] [PubMed]

29. Holt, J.K.; Park, H.G.; Wang, Y.; Stadermann, M.; Artyukhin, A.B.; Grigoropoulos, C.P.; Noy, A.; Bakajin, O. Fast mass transport through sub-2-nanometer carbon nanotubes. *Science* **2006**, *312*, 1034. [CrossRef] [PubMed]

30. Sholl, D.S.; Johnson, J.K. Making high-flux membranes with carbon nanotubes. *Science* **2006**, *312*, 1003–1004. [CrossRef] [PubMed]

31. Chen, X.; Cao, G.; Han, A.; Punyamurtula, V.K.; Liu, L.; Culligan, P.J.; Kim, T.; Qiao, Y. Nanoscale fluid transport: Size and rate effects. *Nano Lett.* **2008**, *8*, 2988. [CrossRef] [PubMed]

32. Bushuev, Y.G.; Sastre, G.; Juliánortiz, J.V.D.; Gálvez, J. Water-hydrophobic zeolite systems. *J. Phys. Chem. C* **2012**, *116*, 24916–24929. [CrossRef]

33. Kumzerov, Y.A.; Nabereznov, A.A.; Vakhrushev, S.B.; Savenko, B.N. Freezing and melting of mercury in porous glass. *Phys. Rev. B Condens. Matter* **1995**, *52*, 4772–4774. [CrossRef] [PubMed]

Appl. Sci. **2018**, *8*, 1065

34. John A, T. Water flow in carbon nanotubes: Transition to subcontinuum transport. *Phys. Rev. Lett.* **2009**, *102*, 184502.
35. Rigby, S.P.; Edler, K.J. The influence of mercury contact angle, surface tension, and retraction mechanism on the interpretation of mercury porosimetry data. *J. Colloid Interface Sci.* **2002**, *250*, 175–190. [CrossRef] [PubMed]

© 2018 by the authors. Licensee MDPI, Basel, Switzerland. This article is an open access article distributed under the terms and conditions of the Creative Commons Attribution (CC BY) license (http://creativecommons.org/licenses/by/4.0/).

![applied sciences logo] *applied sciences*

MDPI

Article

Silica Pillared Montmorillonites as Possible Adsorbents of Antibiotics from Water Media

Maria Eugenia Roca Jalil [1,2,*], Florencia Toschi [1], Miria Baschini [1,2] and Karim Sapag [3]

[1] PROBIEN (CONICET-UNCo), Buenos Aires 1400, Neuquén 8300, Argentina; florencia.toschi@probien.gob.ar (F.T.); miria.baschini@fain.uncoma.edu.ar (M.B.)

[2] Dpto de Qca-Facultad de Ingeniería, Universidad Nacional del Comahue, Buenos Aires, Neuquén 1400, Argentina

[3] Laboratorio de Sólidos Porosos, Instituto de Física Aplicada–CONICET, Universidad Nacional de San Luis, Ejército de los Andes 950, Bloque II, 2do piso, San Luis 5700, Argentina; sapag@unsl.edu.ar

* Correspondence: eugenia.rocajalil@probien.gob.ar; Tel.: +54-2994490300 (ext. 688)

Received: 18 July 2018; Accepted: 16 August 2018; Published: 19 August 2018

Featured Application: Silica pillared clays have not been studied as adsorbents from water media. However, Si-PILC represent an interesting option as selective adsorbents of emerging compounds.

Abstract: In this work, three silica pillared clays (Si-PILC) were synthetized, characterized, and evaluated as possible adsorbents of ciprofloxacin (CPX) and tetracycline (TC) form alkaline aqueous media. The pillared clays obtained showed significant increases in their specific surface areas (S_{BET}) and micropore volumes ($V_{\mu p}$) regarding the raw material, resulting in microporosity percentages higher than 57% in all materials. The studies of CPX and TC removal using pillared clays were compared with the natural clay and showed that the Si-PILC adsorption capacities have a strong relationship with their porous structures. The highest adsorption capacities were obtained for CPX on Si-PILC due to the lower molecular size of CPX respect to the TC molecule, favoring the interaction between the CPX^- and the pillars adsorption sites. Tetracycline adsorption on silica pillared clays evidenced that for this molecule the porous structure limits the interaction between the TCH^- and the pillars, decreasing their adsorption capacities. However, the results obtained for both antibiotics suggested that their negative species interact with adsorption sites on the pillared structure by adsorption mechanisms that involve inner-sphere complex formation as well as van der Waals interactions. The adsorption mechanism proposed for the anionic species on Si-PILC could be considered mainly as negative cooperative phenomena where firstly there is a hydrophobic effect followed by other interactions, such as der Waals or inner-sphere complex formation.

Keywords: silica pillared clays; antibiotics adsorption

---

1. Introduction

The presence of antibiotics in the environment has been widely reported in the last decades and it is a key indicator of the low efficiency of the current effluent treatments in their complete removal. Therefore, several studies have focused on developing techniques to be used as a complement to the actual treatments, allowing the complete removal of this kind of contaminant. Due to the operational simplicity, simple design, and availability of several adsorbents, adsorption is one of the most interesting studies [1–5]. Many adsorbents have been proved to remove antibiotics from water media [6–10]. Among them, clay minerals have low cost and are environmentally compatible materials and, bentonites in particular, have shown to be good adsorbents of many different pharmaceutical compounds [11–17]. Nevertheless, the behavior of bentonite suspensions produces some limitations in their separation from the adsorption aqueous medium and a possible solution for this is to modify the natural clay with the aim of increasing its hydrophobicity. From bentonites is possible to synthesize micromesoporous materials with high specific surface areas and permanent porosities regarding raw materials [17]. These materials, named pillared clays (PILC), have shown to be good adsorbents for different organic compounds with the advantage of having higher hydrophobicities than natural clays, which allows for an easier separation of the adsorption media [18–25].

The pillared clays usually studied as adsorbents are aluminum, zirconia, and iron pillared clays. In a previous work, four different PILC were synthetized and evaluated as ciprofloxacin adsorbents from aqueous media [25]. Among them, one silica pillared clay showed to be a good adsorbent of this antibiotic and the results suggest that there exists an interesting relationship between the porous structure of the Si-PILC and its adsorption capacity. Up to now, the silica pillared clays have been used mainly as catalysts but there are no studies about their adsorbent capacities. Therefore, in this work, three silica pillared clays were synthetized and characterized by different techniques with the aim of studying the relationship between their porous structure and their removal capacity for ciprofloxacin and tetracycline from aqueous media.

2. Materials and Methods

2.1. Synthesis and Characterization of Silica Pillared Clays (Si-PILC)

The raw material used in this work was bentonite (natural clay, NC) from Pellegrini Lake in the province of Rio Negro, Argentina, described in detail in a previous work [16]. The silica pillared clays (Si-PILC) were prepared following the methodology described by Han et al. [26] with some modifications. The silica sol solution used as pillaring agent was obtained by mixing tetraethyl orthosilicate (TEOS: $Si(OEt)_4$, Merck > 99%), 2 M HCl (Cicarelli, 36.5–38%), and ethanol in a molar ratio of 1:0.1:1. The resulting solution was then aged at room temperature and mixed with a 0.25 M ferric nitrate ($Fe(NO_3)_3.9H_2O$, Anedra 99%) solution in a molar ratio of Si/Fe 10:1. This mixture was titrated with a 0.2 M NaOH (Anedra, 98%) solution up to a pH of 2.7. Then, the pillaring agent obtained was added drop-wise to an NC suspension of 1 wt % deionized water and, considering its cation exchange capacity (CEC), the molar ratio chosen was Si:Fe:CEC 50:5:1, respectively. During the cation exchange, the mixture was stirred 3 h at 60 °C and the solid was separated by centrifugation at 3500 rpm for 15 min by a Sorvall RC 5C centrifuge. This solid was washed with a solution of ethanol/water 50% v/v and then was dispersed in a 0.2 M HCl solution under stirring for 3 h. This last step was carried out four more times. Finally, the solid material was washed with deionized water several times and then dried to finally obtain the Si-pillared clay precursor. Three materials were synthetized by this procedure varying the molar ratio between the Si and natural clay CEC and obtaining Si-PILC precursors with ratios of 25, 50, and 75 mol $Si.g^{-1}$ of natural clay. Finally, the precursors of the Si-PILC were calcined at 500 °C for 1 h in order to obtain the Si-PILC 25, 50, and 75, respectively.

The structural properties of the materials were analyzed by X-Ray Diffraction (XRD) and Fourier transform infrared spectroscopy (FTIR). The XRD were obtained using a RIGAKU Geigerflex X-ray diffractometer with CuKα radiation at 20 mA and 40 kV. The scans were recorded between 2° and 70° (2θ) with a step size of 0.02° and a scanning speed of 2° min^{-1}. The FTIR spectra were acquired using an Infralum FT-08 from 300 to 4000 cm^{-1}, the samples were prepared by potassium bromide pressed disc technique (3 mg of sample with 300 mg of KBr). The textural properties were studied by nitrogen adsorption–desorption isotherms at −196 °C. These measurements were carried out using an Autosorb 1MP and iQ (Quantachrome Instruments). All the samples were previously degassed for 12 h up to a residual pressure lower than 0.5 Pa at 200 °C. Textural properties were obtained from these isotherms by different methods. The specific surface area (S_{BET}) was assessed by the Brunauer, Emmet, and Teller (BET) method, using the Rouquerol's criteria [27]. The micropore volumes ($V_{\mu p}$) were calculated with the α-plot method using the corresponding sample calcined at 1000 °C as reference material [28]. The total pore volume (V_T) was obtained using the Gurvich rule (at 0.97 of relative pressure) [27].

Pore size distributions (PSD) were obtained by the Horvarth–Kawazoe method, considering the adsorption branch and that the PILC have slit shape pores within the interlayer region.

2.2. Adsorptives: Ciprofloxacin and Tetracycline

The antibiotics used in this study were ciprofloxacin hydrochloride (CPX, Romikim S.A) and tetracycline hydrochloride (TC, Parafarm). The structures and general characteristics of these two organic compounds are presented in Table 1. Both antibiotics have polifunctional molecules, which implicates that different species are present in the solution depending on the pH media. The CPX molecule has two pKa values giving three ionic species, the cationic (CPX$^+$), the zwiterionic (CPX$^\pm$), and the anionic (CPX$^-$). On the other hand, the TC molecule has three protonable-deprotonable groups associated with their pKa values, giving four different ionic species, the cationic (TCH$_3^+$), the zwiterionic (TCH$_2^\pm$), and two anionics (TCH$^-$ and TC^{2-}). The distribution of antibiotics species at different pH values were obtained by the method reported by Del Piero et al. [29].

Table 1. General characteristics of the antibiotics used.

	Ciprofloxacin.HCl	Tetracycline.HCl
Structure		
Chemical formula	C$_{17}$H$_{18}$FN$_3$O$_3$.HCl	C$_{22}$H$_{24}$N$_2$O$_8$.HCl
Molecular weight (g.mol^{-1})	366.80	480.90
Molecular dimensions (nm)	$1.2 \times 0.3 \times 0.7$ [30]	$1.2 \times 0.8 \times 1.3$ [31]
pKa	pKa$_1$ = 5.9; pKa$_2$ = 8.9	pKa$_1$ = 3.3; pKa$_2$ = 7.7; pKa$_3$ = 9.7

The protonation-deprotonation reactions that take place at different pH values of the media also affect the solubility of the antibiotics. In a previous work [16], the solubility of CPX as a function of the complete pH range was studied. In this work, the solubility curve of TC was obtained from different saturated solutions adjusted at pH values between 2 and 13 (using HCl or NaOH, respectively) and kept at 20 °C for about 12 h (considering that the equilibrium time was reached). After that, the solutions were filtered and the TC concentration was measured by UV-Vis spectroscopy at the corresponding wavelength.

2.3. Adsorption Studies

The adsorption experiments were conducted by mixing 0.02 g of the adsorbent with 8 mL of antibiotic solution in tubes of 10 mL and under stirring at 20 °C until equilibrium was reached. The values were chosen according to the solubility results. In all the tests, the tubes were wrapped in aluminum foils to prevent light-induced decomposition. After adsorption, the solutions were separated from the adsorbent using a Sorvall RC 5C centrifuge at 8000 rpm for 20 min. The antibiotic equilibrium concentrations in the resultant supernatant were measured with a T60 UV-vis spectrophotometer at $\lambda_{máx}$ corresponding to the pH value, from a previously determined calibration curve. All samples were measured in duplicate and the average value was used as isotherm data. The amount of antibiotic adsorbed on the clay mineral (q) was calculated from the initial and equilibrium concentrations, according to Equation (1),

$$q = \frac{(C_i - C_e) * V}{w} \tag{1}$$

where V is the antibiotic solution volume (L), C_i is the initial antibiotic concentration (mg.L^{-1}), C_e is the equilibrium antibiotic concentration (mg.L^{-1}), and w is the mass of clay (g).

2.4. Modelling Methods

Adsorption kinetics of antibiotics on NC and Si-PILC were performed using an initial fixed antibiotic concentration (110 mg.L^{-1} for CPX and 480 mg.L^{-1} TC, respectively) with contact times varying between 0.5 and 24 h. The kinetics studies were useful to assess the contact time required to reach equilibrium. In order to acquire additional information about adsorption mechanisms, experimental data were fitted to the pseudo-first-order, pseudo-second-order, and intraparticle diffusion models [16,18,32,33].

The Lagergren pseudo-first order equation can be expressed as (Equation (2)):

$$q_t = q_e(1 - e^{(-k_1 t)}) \tag{2}$$

where q_t is the amount of antibiotic adsorbed at time t (mg.g^{-1}), q_e is the equilibrium adsorption capacity of the adsorbent (mg.g^{-1}), and k_1 is the rate constant of pseudo first-order kinetics (min^{-1}).

Similarly, if the mechanism is thought to be pseudo-second order kinetic, the equation can be expressed as (Equation (3)):

$$q_t = \frac{k_2 q_e^2 t}{1 + k_2 q_e t} \tag{3}$$

where q_t is the amount of antibiotic adsorbed at time t (mg.g^{-1}), q_e is the equilibrium adsorption capacity of the adsorbent (mg.g^{-1}), and k_2 is the rate constant of pseudo second-order kinetics (g.mg.min^{-1}).

The intraparticle diffusion model was derived from Fick's second law of diffusion and assumes that intraparticle diffusion is the rate limiting step, the film diffusion can be negligible at the beginning of diffusion and the intraparticle diffusivity is constant. The equation of the intraparticle diffusion model can be expressed as it is shown in Equation (4). Thus, the plot of q_t versus $t^{0.5}$ can give one straight line or shows multilinearity which can be related to different diffusion mechanisms [18,33].

$$q_t = k_d \, t^{0.5} \tag{4}$$

where q_t is the amount adsorbed (mg.g^{-1}) at time t and k_d is the intraparticle-diffusion parameter (mg.g^{-1}min$^{-0.5}$).

The adsorption equilibrium data were fitted to Langmuir, Freundlich, and Sips isotherm models [34]. The Langmuir model assumes one adsorption energetically equivalent without interactions among the adsorbed neighbor molecules on a homogeneous surface. The mathematical expression of the Langmuir model is shown in Equation (5):

$$q = \frac{q_m k C_{eq}}{1 + k C_{eq}} \tag{5}$$

where q_m is the maximum adsorbed amount within a monolayer of adsorptive (mg.g^{-1}), k (mg^{-1}.L) is the Langmuir dissociation constant, which is related to the adsorption energy, and C_{eq} has the same meaning as above.

The Freundlich equation is commonly used for multisite adsorption isotherms, it is an empirical method related to adsorption on heterogeneous surfaces and its mathematical expression is (Equation (6)):

$$q = k_f C_e^{1/n} \tag{6}$$

where k_f (L.g^{-1}) and n (dimensionless) are the Freundlich characteristic constants, indicating the adsorption capacity and adsorption intensity, respectively.

The third isotherm model is considered as a combination of the Langmuir and Freundlich equations and is called the Sips model. It is also an empirical equation, related to more heterogeneous adsorption systems without adsorbate–adsorbate interactions; its mathematical expression is shown in Equation (7).

$$q = q_m \frac{(bC_e)^{1/n}}{1 + (bC_e)^{1/n}} \tag{7}$$

where q_m and C_e have the same meanings as above, b is a parameter related to the affinity of the adsorbate towards the surface, and n is a parameter that represents the heterogeneity of the system.

The Scatchard plots were built by transformation of the isotherm data to obtain a plot of q/C_e versus q (where q and C_e have the same meanings indicated above). The shape of the Scatchard plot can be useful to acquire complementary information about the adsorption phenomena [25,35,36].

3. Results and Discussion

3.1. Adsorbents Characterization

XRD patterns obtained for the natural and Si-PILC clays and their main fractions are shown in Figure 1A. The obtained XRD pattern for the NC exhibits a basal distance (d_{001}) of 1.26 nm, which is typical of natural sodic montmorillonite. The Si-PILC diffractograms did not show any defined peaks in the range in which the basal distance should be present. Furthermore, there are no other structural changes in the PILC in reference to the raw material [37,38]. In order to evaluate the existence of an increase in the basal distance (d_{001}), XRD at low angle was obtained for the Si-PILC 75 and is shown in Figure 1B. A basal distance of 4.32 nm was found for this material, which evidences the presence of silica oxide species within the interlayer clay suggesting an effective pillarization process. This result agrees with the values reported by other authors [26]. Si-PILC 25 and 50 were not analyzed at low angle but a similar behavior could expect for them.

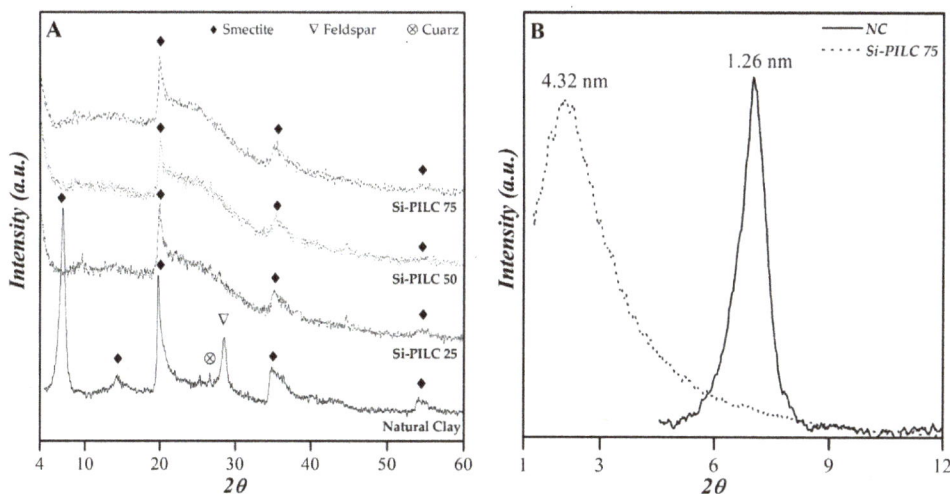

Figure 1. XRD patterns of NC and Si-PILC. (XRD: X-Ray Diffraction.)

The FTIR spectra obtained for the four materials and their main bands are shown in Figure 2. The natural clay spectrum is in accordance with previous reports, where the absorption bands at 3630, 3454, and 1640 cm^{-1} are typical for smectites and associated to tension vibrations of the O–H bond in hydroxyl groups of dioctahedral structures and H–O–H vibrations of water molecules weakly hydrogen bonded to the Si–O surface, respectively [26,39]. In addition, the bands found at 1040, 800, 524, and 470 cm^{-1} are attributed to Si–O stretching vibrations, Si–O–Al (octahedral Al) and Si–O–Si bending vibrations, respectively. The spectra obtained for the Si-PILC show a considerable reduction of the bands related to water presence and O–H vibrations, which could be due to the acid treatment used during the synthesis of these materials [26,39]. Additionally, the spectra of the Si-PILC show increments in the bands of 1200 and 800 cm^{-1} which evidences the increase of Si–O–Si bonds in their structures as a consequence of the pillaring process.

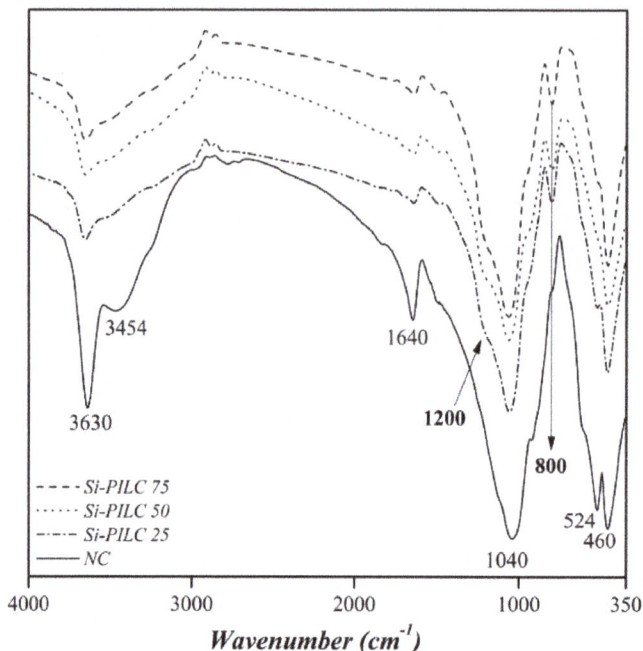

Figure 2. FTIR spectra for the natural and Si-pillared clays.

Nitrogen adsorption–desorption isotherms at −196 °C obtained from the materials are shown in Figure 3. The natural clay isotherm is classified as type IIb with an H3 hysteresis loop according to the International Pure and Applied Chemistry (IUPAC) classification which is associated with mesoporous materials with aggregates of plate-like like the montmorillonites [27,40]. The isotherms obtained for Si-PILC can be classified as a combination between type I and IIb. The first one is due to the high amount adsorbed at low relative pressures related to the presence of microporosity in the PILC structures generated after the pillaring process. The second one is associated with the growth of the adsorption in the mono-multilayer region.

Hysteresis loops were observed for the three Si-PILC demonstrating the presence of mesoporosity in their structures. The type of hysteresis loop is classified as type H4 according to the IUPAC and it is associated with slit-like pores as well as the pores generated within the interlayer of the clay minerals [27,40]. As can be seen, all the pillared clays present higher adsorption at low relative pressures than the natural clay, which is associated with their microporosity and evidences a successful pillaring process. However, among the pillared clays Si-PILC 50 showed the highest adsorption at low relative pressure suggesting there is not a correlation between the adsorption increase observed and the Si/clay ratio used during the synthesis process. This could be due to the amount of siliceous present in the interlayer for the Si-PILC 75 which could generate higher pillars density than the Si-PILC 50 implicating a lower microporosity.

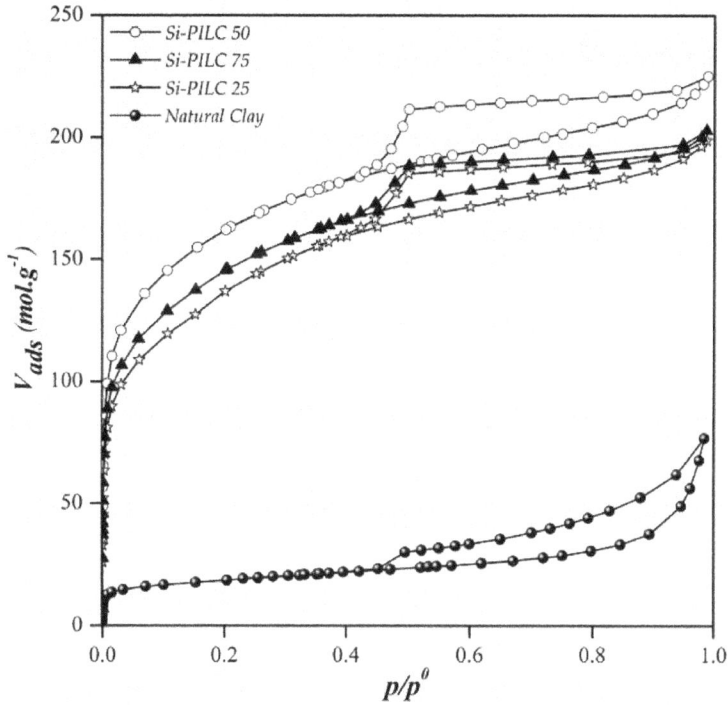

Figure 3. N_2 adsorption–desorption isotherms for the four materials.

Textural properties for all materials were obtained from the N_2 adsorption–desorption isotherms and are summarized in Table 2. Significant increases in the specific surface areas (S_{BET}) for the Si-PILC with regard to the natural clay are observed, reaching nine times the NC value for the Si-PILC 50. This last is related to the increase in the microporosity of these materials resulting in 57, 65, and 61% for the Si-PILC 25, 50, and 75, respectively. The S_{BET} and micropores volume ($V_{\mu p}$) values increased from Si-PILC 25 to Si-PILC 75 and Si-PILC 50, respectively, and it is associated with the amount of micropores present in their structures. The fact that Si-PILC 75 showed lower values than Si-PILC 50 could be due to the amount of pillaring agent added. The results suggest that Si/clay ratio higher than 50 mol Si.g^{-1} could generate more pillars density within the clay interlayer which during the heating treatment could collapse, decreasing the microporosity. The V_T values supported the behavior of the isotherms at high relative pressures obtaining the highest value for the Si-PILC 50.

Table 2. Textural properties for the materials. (PILC: pillared clays.)

	S_{BET} (m^2.g^{-1})	V_T (cm^3.g^{-1})	$V_{\mu p}$ (cm^3.g^{-1})
Natural Clay	67	0.10	0.01
Si-PILC 25	485	0.30	0.17
Si-PILC 50	585	0.34	0.22
Si-PILC 75	519	0.31	0.19

The PSD for the PILC and NC were studied in the micropores region (pores size below 2 nm) and are shown in Figure 4. As can be observed, the microporosity in the natural clay is considerably lower than that obtained for the pillared clay, which is in accordance with the textural results.

All pillared clays have micropore sizes between 0.5 and 2 nm, whereas Si-PILC 25 and 50 showed similar distributions with micropore sizes approximately 0.75 nm. Si-PILC 75 shows contribution of smaller pore sizes (near to 0.7 nm) than the other PILC, suggesting the greater pillar density within the interlayer due to the high Si/clay ratio.

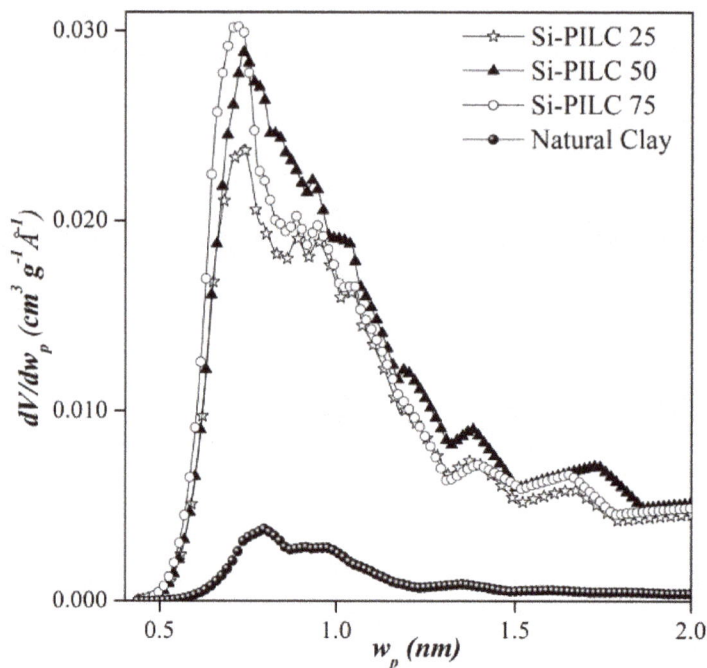

Figure 4. Pore size distribution (PSD) of natural and pillared clay minerals.

3.2. Effect of the pH Media on the Adsorption

The presence of protonable groups in the antibiotics molecules generates different species according to the media pH. Thus, the species present in the aqueous solution and their solubility depend on the pH; this also influences their interaction with the adsorbent surfaces. In a previous work, the strong correlation between the solubility in water and the adsorption on clay of CPX with the pH of the media was shown [16]. The tetracycline solubility curves in water with its distribution of species as a function of the media pH are shown in Figure 5. The solubility behavior can be explained considering the species present within each pH range. As was mentioned above, the TC molecule has three pK_a values which implicate the existence of four possible species in solution. At the lowest pH values the soluble cationic TC species (TCH_3^+) is present and its percentage decreases from pH 2 to 3.3 (pK_{a1}) as well as its solubility. In the pH range from 3.3 to 7.7 (pK_{a2}) three TC species exist, where the zwitterion form ($TCH_2^\pm$) is the predominant one, reaching the lowest solubility due to its neutral charge. Later increases in the media pH generate the increase in solubility because TC turns into its anionic forms, TCH^- (between pH values of 7.7 and 9.7) and TC^{2-} (pH values higher than 9.7). Thus, analogously to the results for CPX, TC is more soluble when the ionic forms are present in the media and shows the lowest solubility around a pH value of 4.8 because at this pH it has the highest zwitterion percentage. Additionally, it is known that due to the existence of asymmetric carbon atoms in the TC molecule, its species in solution can adopt two possible conformations called the extended and twisted conformation [41]. The first one is predominant in basic media, whereas the second one in neutral or acid media. This must be taken into consideration because according to the conformation the molecular size can suffer changes.

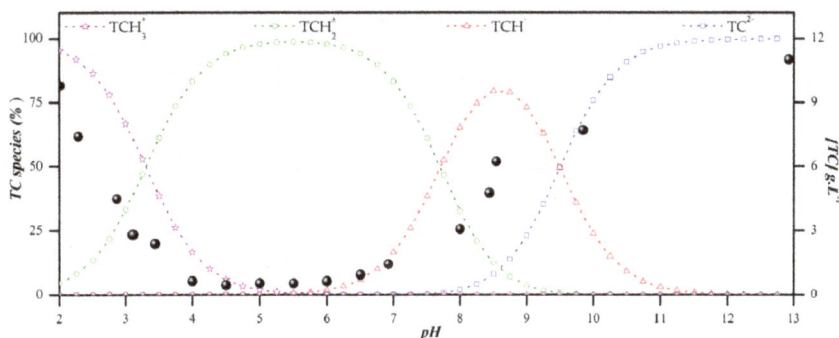

Figure 5. TC solubility curve and distribution of species as function of pH media.

The studies of antibiotic adsorption on NC and Si-PILC 75 at different pH values were assessed under the above mentioned conditions with an initial fixed concentration of 110 and 480 mg.L^{-1} for CPX and TC, respectively. The pH of the initial solution was adjusted to values between 3 and 10 using HCl or NaOH solutions. These values were chosen based on the antibiotics solubility curves and considering the species present in each pH range [16]. The adsorption capacity of CPX and TC on NC and Si-PILC 75 at different pH values are shown in Figure 6. In the case of TC, its adsorption on NC showed a behavior similar to the reported by other authors, where the highest adsorption occurs at low pH values, when the TCH$_3^+$ species is present and the adsorption is governed mainly by electrostatic interactions [42]. The cationic form of TC is adsorbed on the negative charge surface of the NC mainly by cation exchange for the natural cation within the montmorillonite interlayer [42,43]. After that, at pH values greater than 4 the adsorption decreases with increases in pH because the TC transitions into zwitterionic and anionic forms. At these pH values, the species could be adsorbed on the mineral negative surface by other adsorption mechanisms like hydrophobic interactions, hydrogen bonding, or inner sphere complexes. However, it is important to consider that at pH values lower than 10, TCH$_2^\pm$ and TCH$^-$ species have net charges of zero and one, respectively, but the amine group is protonated in their structures. This could favor the adsorption of these species on the negatively charged surface of the natural clay. Other authors proposed that when these species are adsorbed within the interlayer clay, they arrange at the surface so that the positively charged group is located close to the surface sites and their negatively charged part is driven as far away as possible from the surface [42,43]. The lowest TC adsorption was obtained at pH values higher than 9.7, where the main species was the TC^{2-} and it was probably because of the repulsion with the negatively charged surface of the NC. The TC adsorption on Si-PILC 75 was considerably lower than that obtained for the NC in all the pH range except to values up to 8.5. No greater differences in the amount of TC adsorbed on pillared clay were observed in the whole pH range. This could suggest that the interactions between the TC species and the pillared clay surface are low, or that the porous structure of this material could be avoiding the molecule access to the adsorption sites. The adsorption of CPX on natural clay showed a similar behavior to the TC adsorption, were the highest adsorptions were obtained at low pH values with a maximum at pH 6; decreasing when the pH increases. These results can be explained similarly to those above; at low pH values the CPX$^+$ is present and is adsorbed on the negative surface of the natural clay by cation exchange. After that, at pH values of 7.5, CPX$^\pm$ and CPX$^-$ are present in the solution, decreasing the adsorption and resulting in other adsorption mechanisms [16]. Results shown for the adsorption of CPX on Si-PILC 75 evidenced that there are no variations in the amount adsorbed at different pH values, decreasing with an increase in the pH. However, the adsorption for this material under basic conditions (up to 7.5 pH) was higher than for the natural clay, suggesting that Si-PILC interacts with the CPX species by another mechanism, like the formation of an inner sphere complex [25,44].

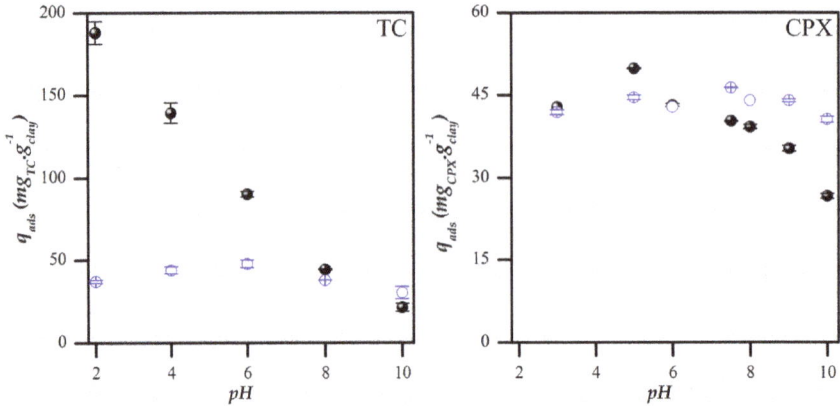

Figure 6. Effect of media pH on CPX and TC adsorption where the full and empty symbols are NC (●) and Si-PILC 75 (○), respectively.

3.3. Adsorption Kinetics

Adsorption kinetics of both antibiotics on NC and Si-PILC 75 were evaluated at pH values of 8.5 and 10 for TC and CPX, respectively, so that the species in solution were anionic in order to evaluate their interactions with the surfaces materials. The experiments were performed using the same initial and fixed concentrations than above, 110 and 480 mg.L^{-1} for CPX and TC with contact times varying between 0.5 and 24 h. Adsorption kinetic curves obtained for the Si-PILC 75 and their fitting to the pseudo-first and pseudo-second order are shown in Figure 7. For the adsorption on NC, the results showed that the adsorption equilibrium is rapidly reached for TC and is lower for CPX (around 4 h). These results could be due to the species present in each case. For the TC at a pH value of 8.5, the species present is TCH^{-} and its structure still has a positively charged group. This last favors the adsorption on the negative surface of the NC by cation exchange, implicating a fast adsorption. In the case of CPX at a pH value of 10, the species present is anionic and the higher time to reach the equilibrium could be due to the fact that the adsorption mechanism is not cation exchange. No significant differences were obtained among the data, indicating that the equilibrium of the systems was reached, and therefore the kinetic studies on NC were not fitted to the models. The TC adsorption on Si-PILC 75 reached equilibrium at around 7 h, what could be due to the fact that the pillared clay porous structure is limiting the access of the TC molecule to the interlayer region and delaying the adsorption. This implicates that the adsorption mechanism in the Si-PILC is different from the one proposed for the NC. The results obtained for CPX on Si-PILC 75 are similar to those obtained for TC, the system reaches the equilibrium at a contact time of 8 h. This supports the fact that in the pillared clay its microporous structure could be avoiding the access of the molecules to the interlayer region generating a slower adsorption process.

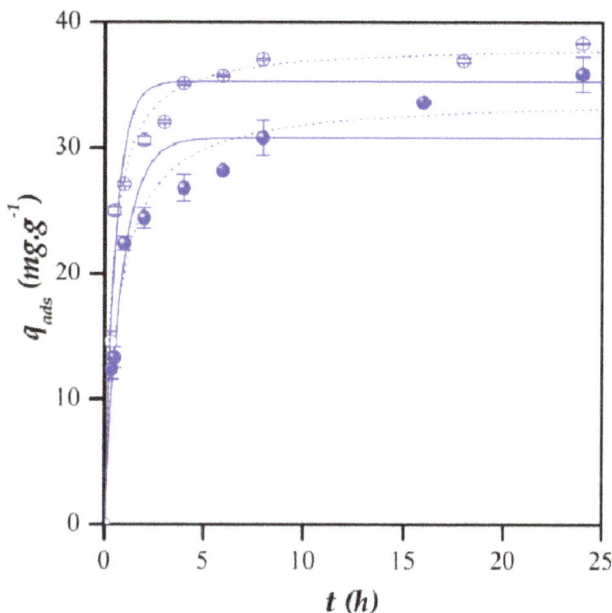

Figure 7. Kinetic adsorption data for TC (●) and CPX (O) on Si-PILC 75. Solid lines represent a pseudo-first-order model and dot lines represent pseudo-second order model.

The adsorption kinetic parameters were estimated by nonlinear regression of pseudo-first and pseudo-second order equations and for two materials [32,45]. The obtained values are in good agreement with the previously shown results obtained for the adsorption vs. pH curves and are summarized in Table 3. As can be seen from the table, all the regression coefficients are higher than 0.90. However, if the studied systems are considered, the second-order model is more appropriate due to the fact that it is based on the sorption capacity of the solid and it considers that the rate controlling step is the chemical or electrostatic change generated in the system [32,44,45]. In addition, the k_2 values agree with the fact that natural clay requires lower time than the pillared clay to reach equilibrium in both systems. For the kinetics adsorption results shown, an equilibrium time of 24 h was chosen to perform the adsorption isotherms.

In order to acquire more information regarding the adsorption process on the pillared clay, the intraparticle model was used and the results obtained are shown in Figure 8. As can be seen from this figure, the plots resulted in two different straight lines with regressions that do not pass through the origin, indicating that intraparticle diffusion was not the only rate controlling step. This could be related to the existence of an initial external mass transfer or a chemical reaction between the antibiotic molecule and the adsorbent surface [33]. The slopes of the linear regressions (k'_3 and k''_3 in Table 3) indicate the rate of the adsorption process. The values obtained for the intraparticle diffusion parameters indicate the decrease of the diffusion rates with the increase of the contact time for both antibiotics. These results evidenced that as the antibiotic molecules diffused into the pillared clay structure, the pores aviable for diffusion decreased, limiting the adsorption. Thus, the first section is related to the stage in which the antibiotics are rapidly adsorbed on the PILC surface, which could be related to the microporous structure. The second section is related to the adsorption on the external surface, which contains the mesoporous structure. Similar results were reported for the adsorption of dyes on pillared clays [18].

Table 3. Pseudo-first-, -second-order, and intraparticle rate parameters for TC and CPX on Si-PILC 75.

		TC	CPX
		pH 8.5	pH 10
Pseudo first order	q_e (mg.g^{-1})	30.8	35.2
	k_1 (1 min^{-1})	1.10	1.86
	R^2	0.93	0.93
Pseudo second order	q_e (mg.g^{-1})	33.9	3.21
	k_2 (g(mg.min)$^{-1}$)	0.04	0.07
	R^2	0.98	0.98
Intraparticle	k'_3 (mg.(g.min$^{0.5}$)$^{-1}$)	1.91	0.92
	R^2	0.99	0.99
	k''_3 (mg.(g.min$^{0.5}$)$^{-1}$)	0.40	0.13
	R^2	0.97	0.95

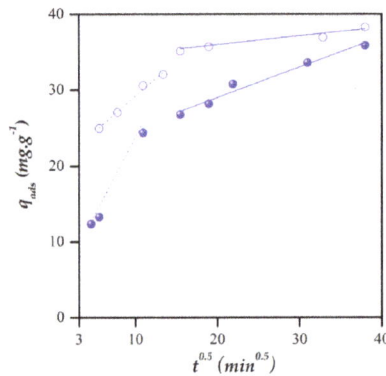

Figure 8. Intraparticle diffusion model for the adsorption of TC (●) and CPX (O) on Si-PILC 75.

3.4. Adsorption Isotherms

The batch adsorption experiments were performed in the conditions mentioned above varying the initial concentrations between 40–610 mg.L^{-1} for the TC and 18–500 mg.L^{-1} for the CPX. The contact time was set to a period of 24 h and the contact solutions were set at pH 8.5 and 10 for TC and CPX, respectively. These values were chosen based on the results previously shown and in order to evaluate the behavior of the clays adsorption when the species with a net charge of -1 are present (CPX$^-$ and TCH$^-$). The adsorption isotherms obtained for the antibiotics on four materials and their best adjustments are shown in Figure 9. Taking into account the Giles et al. classification [46], different behaviors can be associated with the isotherm shapes obtained. For the adsorption of TC on the natural clay, the isotherm can be classified as sigmoidal type (S type) which is considered to be a result of two opposite mechanism. In first place the TCH$^-$ species has a low affinity towards the negative surface of the clay but once some molecules are adsorbed, they can become new adsorption sites favoring subsequent adsorptions. This phenomenon is called cooperative adsorption and the isotherm saturation point occurs when no vacant sites remain [46,47]. The isotherms obtained for the adsorption of TC on pillared clays showed some differences among them. The Si-PILC 50 and 75 isotherms can be classified as high affinity type (H-type) and the one obtained for the Si-PILC 25 was a Langmuir type (L-type). Both types of isotherms are obtained when the saturation of solid sites is progressive until reaching its limited adsorption capacity [46,47]. The H-type isotherm is usually associated with the adsorption of ionic solutes, where the competition between adsorptive and solvent molecules is low. This could be the result of the higher hydrophobicity exhibited by the pillared clays in contrast with the natural clay. Also, these results evidenced that pillared clays have more affinity for the TCH$^-$ species present in solution than natural clay. The high affinity shown suggests that the adsorption occurs by a different mechanism and it could be due to the presence of new adsorption sites on the surface of the pillared clays associated with the pillars. Additionally, Si-PILC 25 was the pillared clay with the lowest affinity but the highest adsorption capacity and this could be explained considering that it was the material with the lowest ratio Si/clay. This could implicate that the Si-PILC 25 interlayer region has probably lower pillars density resulting in a porous structure that favors the interaction between the TC species and the pillars. The adsorption isotherms obtained for CPX can be classified as H-type for Si-PILC 50 and 25 and L-type for Si-PILC 75 and NC. Their analysis is similar to the one done for TC adsorption, all the pillared clays showed a higher affinity towards the CPX$^-$ than the natural clay. The natural clay isotherm suggests a lower affinity of the anionic CPX species present towards the more negatively charged clay than the one observed for the PILC. As in the case of the TC adsorption, this last could be due to a combination of the adsorption mechanism and the hydrophobicity of these materials.

As can be seen in Figure 9, all the materials showed higher CPX adsorption capacities than those obtained for TC adsorption. The adsorption results obtained for the NC evidences that it has higher affinity and adsorption capacity for the CPX$^-$ than for the TCH$^-$, even when these species still have a protonated group in their structure. These results may suggest that these two species are adsorbed in different ways on the natural clay, where TCH$^-$ is probably adsorbed within the interlayer region, whereas the CPX$^-$ interacts with adsorption sites on the clay surface. Further work is required to stablish this idea.

In the case of the Si-PILC, the results evidenced that these materials have a considerably higher affinity and adsorption capacities for the CPX species than for the TC which could be related to the access that the molecules have to the porous structures. In a previous work, the adsorption of CPX on different pillared clays in alkaline media was proposed as result of a combination of two effects. The first one was the porous structure of the PILC limiting the molecule access to the pillared structure, and the other one was the adsorption of the anionic species by inner sphere complex formation with the pillars' adsorption sites [25]. Taking into account the criteria proposed in that work about the porous size needed for the molecules to access to the pillared structure, and considering that the molecules diffuse into the porous structure of the adsorbents lengthwise, the adsorption results could be explained. At alkaline pH, TC shows its extended conformation with dimensions of $1.2 \times 0.8 \times 1.3$ nm which implicates that the size needed to be adsorbed is 1.4 nm, whereas for the CPX, the size is 1.3 nm. There are no significant differences between these two values but considering the dimensions of CPX ($1.2 \times 0.3 \times 0.7$ nm), its highest adsorption could be explained. Also, this may indicate that the CPX molecule has more access to the pillar structure that favors its interaction with the pillars.

All the adsorption data obtained were fitted to Freundlich, Langmuir, and Sips models and their fitting parameters are summarized in Table 4. Considering the results obtained for TC adsorption on NC and Si-PILC 25, Langmuir and Sips models, respectively, are not shown because they have a higher percentage error associated, suggesting that the adsorption does not agree with the assumptions proposed in those models. As can be seen from the Table 4, the best fittings were obtained for the Sips model in all the systems under study. These results are likely related to the fact that there are more heterogeneous adsorption systems which could be generated for the different adsorption sites on the solid surfaces, the adsorptive species or, a combination of them. Regarding the adsorption on the NC, the heterogeneity could be associated to the hydrophobic effect of the solvent towards the antibiotics molecules generating the adsorption as result of its repulsion against the solvent from the solution [25]. This is in accordance with the obtained values for the b parameter for NC systems which suggest that these systems have the lowest affinity of the adsorbates towards the surface. It could be explained for the species present for each antibiotic which are negatively charged, and its adsorption probably occurs by different adsorption short-range forces such as hydrophobic bonding, hydrogen bridges, steric, or orientation effects. These results are according to those mentioned above. In the case of the antibiotics adsorption on Si-PILC, the hydrophobic effect possibly influences in the same way of the NC, but with the highest n values, suggesting more heterogeneous systems. This higher heterogeneity for the PILC compared to the NC could be related to the new adsorption sites generated by the pillars presence and their porous structures. The results showed that for all the PILC materials, the systems for CPX have higher adsorption capacities than for TC and are also more heterogeneous. This can be related to the molecule size, due to the fact that the CPX size is lower than the TC size, allowing more access for CPX to the porous structure and resulting in higher adsorptions and affinities than the TC species. Finally, regarding the adsorption of TC on Si-PILC, the highest adsorption capacity was obtained for Si-PILC 25, and the best fit was obtained with the Freundlich model, suggesting that for this material the adsorption occurs on heterogeneous sites. This last result could be related to the lower amount of micropores in the Si-PILC 25 structure which could allow more access of the TCH$^-$ species to the adsorption sites.

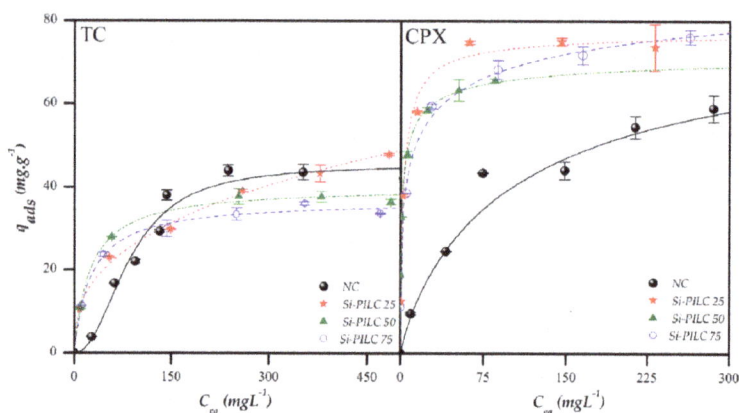

Figure 9. Experimental isotherms (symbols) and their best equations adjustments for the equilibrium adsorption data of TC and CPX on natural and pillared clays.

Table 4. Freundlich and Langmuir parameters for CPX and TC adsorption on natural and pillared clay minerals.

				TC				CPX		
		NC	Si-PILC 25	Si-PILC 50	Si-PILC 75	NC	Si-PILC 25	Si-PILC 50	Si-PILC 75
Freundlich	k_F $(mg \cdot g^{-1}(ppm^{-1})^n)$	3.26	4.81	9.43	8.47	6.98	30.7	31.6	28.9
	n	2.28	2.68	4.36	4.16	2.75	5.52	5.69	5.49
	R^2	0.84	0.99	0.94	0.95	0.96	0.91	0.97	0.98
Langmuir	q_m $(mg \cdot g^{-1})$	—	57.4	40.4	36.7	75.7	74.5	61.9	74.1
	k $((L.mg)^{-1})$	—	0.01	0.04	0.04	0.01	0.43	1.23	0.19
	R^2	—	0.96	0.99	0.99	0.99	0.99	0.98	0.97
Sips	q_m $(mg \cdot g^{-1})$	45.5	—	39.5	36.92	80.82	77.35	73.33	100.6
	b $(L.mg^{-1})$	0.01	—	0.04	0.04	0.01	0.37	0.59	0.07
	n	0.45	—	0.96	1.02	1.13	1.24	1.89	2.56
	R^2	0.97	—	0.99	0.99	0.99	0.99	0.99	0.99

The Scatchard plots obtained for all the materials and both antibiotics are shown in Figure 10. From these plots, the R^2 and binding constant (k_d) values can be obtained. The first could be related to the presence of nonspecific or multitype interactions, while the second indicates the adsorbate affinity for the adsorption sites. The R^2 values obtained for the TC adsorption on four materials were 0.04, 0.78, 0.99, and 0.94 for the NC, Si-PILC 25, 50, and 75, respectively. These values suggest that the nonspecific interactions in the adsorption of TC on pillared clays are considerably higher than for the NC. Similarly, the k_d values obtained were 0.001, 0.02, 0.04, and 0.03 for the NC, Si-PILC 25, 50, and 75, respectively. These values evidence a poor affinity of the TCH$^-$ species for the adsorption sites in the four materials; the lowest affinity was obtained for the NC. In the case of CPX adsorption the values obtained were 0.91, 0.97, 0.93, and 0.74 (R^2) and 0.015, 0.45, 1.36, and 3.07 (k_d) for NC, Si-PILC 25, 50, and 75, respectively. As it can be observed, the presence of multitype interactions is lower for this molecule than for the TC and the k_d values obtained are higher than those obtained for adsorption of TC on all the materials. This could be due to the lower molecular size of CPX favors the interaction with the pillar for the PILC materials.

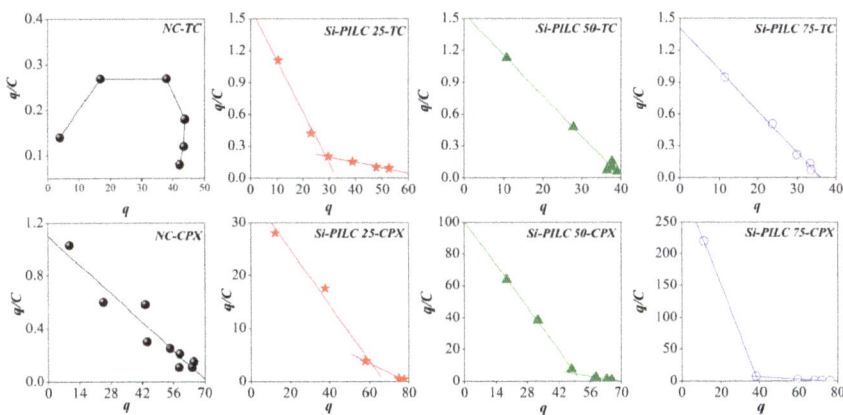

Figure 10. Scatchard plots derived for adsorption data obtained for two antibiotics and the four materials.

Additionally, the Scatchard plot shapes can be related to different adsorption mechanisms. Two different types of plots were obtained for the CPX adsorption, a straight line for the NC and concave curves for the three pillared clays. The first one suggests that there is one type of site for CPX adsorption on the adsorbent surface, whereas the curves acquired for the pillared clays are related to a negative cooperative phenomenon. This last is proposed when two independent sets of data that individually arrange in a linear combination are obtained. Each of them could be associated to two types of affinities of CPX to the surface, one being the consequence of a strong interaction like the formation of inner sphere complexes (high affinity sites) and the other resulting from low affinity interactions like the hydrophobic effect, van der Waals interactions, or hydrogen bridges [25,34,35]. On other hand, Scatchard plots obtained for TC adsorption showed three different shapes. The plot shown for NC resulted in a convex curve suggesting a positive cooperative phenomenon, similar to that proposed for its adsorption isotherms [34,35]. Scatchard plots obtained for the pillared clays showed that Si-PILC 25 has a different behavior than other materials. The plot obtained for TCH$^-$ adsorption on Si-PILC 25 showed a similar behavior that the obtained for the CPX$^-$ on pillared clays, where a negative cooperative phenomenon was proposed as the result of the presence of two different adsorption sites for the adsorbate. However, for Si-PILC 50 and 75, straight lines were obtained suggesting that the adsorption of TCH$^-$ on these materials occurs in the same type of sites. This last could be due to the fact that TC species only has access to one type of adsorption site because the porous structure denies it access to another one. All the results obtained suggest that the adsorption capacity of Si-PILC depends on their porous structure and the adsorbate molecular sizes.

4. Conclusions

In this study, three silica pillared clays were synthetized, characterized, and evaluated as possible adsorbents of antibiotics from alkaline aqueous media. Adsorption data evidenced a strong relationship among the pillared clays' adsorption capacity, their porous structures, and the antibiotics' molecular sizes. The highest adsorption capacities obtained for CPX on Si-PILC are probably due to the fact that its lower molecular size favors access to the pillars adsorption sites. TC adsorption data on silica pillared clays evidenced that for this molecule the porous structure limits the interaction between the TCH$^-$ and the pillars. The adsorption results for both antibiotics suggested that the negative species interact with different adsorption sites on the pillared structure. Finally, the mechanism proposed for the adsorption of anionic species on Si-PILC could be considered as negative cooperative phenomena.

Author Contributions: M.E.R.J. designed and performed the experiments, analyzed the data, and wrote the paper. F.T., M.B., and K.S. cooperated with the data analysis and manuscript elaboration. All the authors read and approved the final manuscript.

Funding: This research was received no external funding.

Acknowledgments: The authors gratefully acknowledge the Universidad Nacional del Comahue, Universidad Nacional San Luis, PROBIEN (CONICET-UNCo), INFAP (CONICET-UNSL), and ANPCyT (Agencia Nacional de Promoción Científica y Tecnológica) for their financial support.

Conflicts of Interest: The authors declare no conflicts of interest.

References

1. Dietrich, D.R.; Hitzfeld, B.C.; O'Brien, E. Toxicology and Risk assessment of pharmaceuticals. In *Organic Pollutants in the Water Cycle*; Reemtsma, T., Jekel, M., Eds.; John Wiley & Sons: Weinheim, Germany, 2006; pp. 287–309.
2. Homem, V.; Santos, L. Degradation and removal methods of antibiotics from aqueous matrices—A review. *J. Environ. Manag.* **2011**, *92*, 2304–2347. [CrossRef] [PubMed]
3. Grassi, M.; Kaykioglu, G.; Belgiorno, V.; Lofrano, G. Removal of Emerging contaminants from water and wastewater by adsorption process. In *Emerging Compounds Removal from Wastewater: Natural and Solar Based Treatments*; Lofrano, G., Ed.; Springer: Dordrecht, The Netherlands, 2012; pp. 15–37.
4. Van Doorslaer, X.; Dewulf, J.; Van Langenhove, H.; Demeestere, K. Fluoroquinolone antibiotics: An emerging class of environmental micropollutants. *Sci. Total Environ.* **2014**, *500*, 250–269. [CrossRef] [PubMed]
5. Akhtar, J.; Amin, N.A.S.; Shahzad, K. A review on removal of pharmaceuticals from water by adsorption. *Desalin. Water Treat.* **2016**, *57*, 12842–12860. [CrossRef]
6. Rakić, V.; Rac, V.; Krmar, M.; Otman, O.; Auroux, A. The adsorption of pharmaceutically active compounds from aqueous solutions onto activated carbons. *J. Hazard Mater.* **2015**, *282*, 141–149.
7. Balarak, D.; Mostafapour, F.K.; Bazrafshan, E.; Saleh, T.A. Studies on the adsorption of amoxicillin on multi-wall carbon nanotubes. *Water Sci. Technol.* **2017**, *75*, 1599–1606. [CrossRef] [PubMed]
8. Liang, Z.; Zhaob, Z.; Sun, T.; Shi, W.; Cui, F. Adsorption of quinolone antibiotics in spherical mesoporous silica: Effects of the retained template and its alkyl chain length. *J. Hazard Mater.* **2016**, *305*, 8–14. [CrossRef] [PubMed]
9. Calisto, V.; Ferreira, C.I.; Oliveira, J.A.; Otero, M.; Esteves, V.I. Adsorptive removal of pharmaceuticals from water by commercial and waste-based carbons. *J. Environ. Manag.* **2015**, *152*, 83–90. [CrossRef] [PubMed]
10. Chen, H.; Gao, B.; Li, H. Removal of sulfamethoxazole and ciprofloxacin from aqueous solutions by graphene oxide. *J. Hazard Mater.* **2015**, *282*, 201–207. [CrossRef] [PubMed]
11. Mabrouki, H.; Akretche, D.E. Diclofenac potassium removal from water by adsorption on natural and Pillared Clay. *Desalin. Water Treat.* **2016**, *57*, 6033–6043. [CrossRef]
12. Wu, H.; Xie, H.; He, G.; Guan, Y.; Zhang, Y. Effects of the pH and anions on the adsorption of tetracycline on iron-montmorillonite. *Appl. Clay Sci.* **2016**, *119*, 161–169. [CrossRef]
13. Al-Khalisy, R.S.; Al-Haidary, A.M.A.; Al-Dujaili, A.H. Aqueous phase adsorption of cephalexin onto bentonite and activated carbon. *Sep. Sci. Technol.* **2010**, *45*, 1286–1294. [CrossRef]
14. Genç, N.; Dogan, E.C.; Yurtsever, M. Bentonite for ciprofloxacin removal from aqueous solution. *Water Sci. Technol.* **2013**, *68*, 848–855. [CrossRef]

15. Jiang, W.-T.; Chang, P.-H.; Wang, Y.-S.; Tsai, Y.; Jean, J.-S.; Li, Z.; Krukowski, K. Removal of ciprofloxacin from water by birnessite. *J. Hazard Mater.* **2013**, *250–251*, 362–369. [CrossRef] [PubMed]

16. Roca Jalil, M.E.; Baschini, M.; Sapag, K. Influence of pH and antibiotic solubility on the removal of ciprofloxacin from aqueous media using montmorillonite. *Appl. Clay Sci.* **2015**, *114*, 69–76. [CrossRef]

17. Gil, A.; Korili, S.A.; Vicente, M.A. Recent Advances in the control and characterization of the porous structure of pillared clay catalysts. *Catal. Rev.* **2008**, *50*, 153–221. [CrossRef]

18. Gil, A.; Assis, F.C.C.; Albeniz, S.; Korili, S.A. Removal of dyes from wastewaters by adsorption on Pillared clays. *Chem. Eng. J.* **2011**, *168*, 1032–1040. [CrossRef]

19. Hou, M.-F.; Ma, C.-X.; Zhang, W.-D.; Tang, X.-Y.; Fan, Y.-N.; Wan, H.-F. Removal of rhodamine B using iron-pillared bentonite. *J. Hazard Mater.* **2011**, *186*, 1118–1123. [CrossRef] [PubMed]

20. Liu, Y.N.; Dong, C.; Wei, H.; Yuan, W.; Li, K. Adsorption of levofloxacin onto an iron-pillared montmorillonite (clay mineral): Kinetics, equilibrium and mechanism. *Appl. Clay Sci.* **2015**, *118*, 301–307. [CrossRef]

21. Mishra, T.; Mahato, D.K. A comparative study on enhanced arsenic (V) and arsenic (III) removal by iron oxide and manganese oxide pillared clays from ground water. *J. Environ. Chem. Eng.* **2016**, *4*, 1224–1230. [CrossRef]

22. Molu, Z.B.; Yurdakoç, K. Preparation and characterization of aluminum Pillared K10 and KSF for adsorption of thimethoprim. *Micropor. Mesopor. Mat.* **2010**, *127*, 50–60. [CrossRef]

23. Roca Jalil, M.E.; Baschini, M.; Rodríguez-Castellón, E.; Infantes-Molina, E.; Sapag, K. Effect of the Al/clay ratio on the thiabendazole removal by aluminum pillared clays. *Appl. Clay Sci.* **2014**, *87*, 245–263. [CrossRef]

24. Roca Jalil, M.E.; Vieria, R.S.; Azevedo, D.; Baschini, M.; Sapag, M. Improvement in the adsorption of thiabendazole by using aluminum pillared clays. *Appl. Clay Sci.* **2013**, *71*, 55–63. [CrossRef]

25. Roca Jalil, M.E.; Baschini, M.; Sapag, K. Removal of Ciprofloxacin from Aqueous Solutions Using Pillared Clays. *Materials* **2017**, *10*, 1345. [CrossRef] [PubMed]

26. Han, Y.S.; Matsumoto, H.; Yamanaka, S. Preparation of new silica sol-based Pillared clays with high Surface area and high thermal stability. *Chem. Mater.* **1997**, *9*, 2013–2018. [CrossRef]

27. Rouquerol, F.; Rouquerol, J.; Sing, K.; Llewellyn, P.; Maurin, G. *Adsorption by Powders and Porous Solids: Principles Methodology and Applications*, 2nd ed.; Elsevier: Amsterdam, The Netherlands, 2013.

28. Villarroel-Rocha, J.; Barrera, D.; García Blanco, A.A.; Roca Jalil, M.E.; Sapag, K. Importance of the α -plot Method in the characterization of nanoporous materials. *Adsorpt. Sci. Technol.* **2013**, *31*, 165–183. [CrossRef]

29. Del Piero, S.; Melchior, A.; Polese, P.; Portanova, R.; Tolazzi, M. A novel multipurpose Excel tool for equilibrium speciation based on newton-raphson method and on a hybrid genetic algorithm. *Ann. Chim.* **2006**, *96*, 29–49. [CrossRef] [PubMed]

30. Bodo, A.; Ciavardini, A.; Giardini, A.; Paladini, A.; Piccirillo, S.; Scuderi, D. Infrered multiple photon dissociation spectroscopy of ciprofloracin: Investigation of the protonation site. *Chem. Phys.* **2012**, *398*, 124–128. [CrossRef]

31. Chopra, I.; Hawkey, P.M.; Hinton, M. Tetracyclines molecular and clinical aspects. *J. Antimicrob. Chemoth.* **1992**, *29*, 245–277. [CrossRef]

32. Ho, Y.S.; McKay, G. A comparison of chemisorption kinetic models applied to pollutant removal on various sorbents. *Process Saf. Environ. Protect.* **1998**, *76*, 332–340. [CrossRef]

33. Wu, F.C.; Tseng, R.L.; Juang, R.S. Initial behavior of intraparticle diffusion model used in the description of adsorption kinetics. *Chem. Eng. J.* **2009**, *153*, 1–8. [CrossRef]

34. Febrianto, J.; Kosasih, A.N.; Sunarso, J.; Ju, Y.; Indraswati, N.; Ismadji, S. Equilibrium and kinetic studies in adsorption of heavy metals using biosorbent: A summary of recent studies. *J. Hazard Mater.* **2009**, *162*, 616–645. [CrossRef] [PubMed]

35. Gezici, O.; Kara, H.; Ayar, A.; Topkafa, M. Sorption behavior of Cu(II) ions on insolubilized humic acid under acidic conditions: An application of Scatchard plot analysis in evaluating the pH dependence of specific and nonspecific bindings. *Sep. Purif. Technol.* **2007**, *55*, 132–139. [CrossRef]

36. Dahlquist, F.W. The Meaning of Scatchard and Hill Plots. In *Methods of Enzymology*; Hirs, C.H.W., Timasheff, S.N., Eds.; Academic Press: New York, NY, USA, 1978; Volume 48, pp. 270–299.

37. Komadel, P.; Madejová, J. Acid activation of Clay Minerals. In *Handbook of Clay Science Developments in Clay Science, Part, A: Fundamentals*, 2nd ed.; Bergaya, F.G., Lagaly, G., Eds.; Elsevier Ltd.: Amsterdam, The Netherlands, 2013; Volume 2, pp. 385–408.

38. Mendioroz, S.; Pajares, A.; Benito, I.; Pesquera, C.; González, F.; Blanco, C. Texture evolution of montmotillonite under progressive acid treatment: Change from H_3 to H_2 type of hysteresis. *Langmuir* **1987**, *3*, 676–681. [CrossRef]

39. Madejová, J. FTIR techniques in clay mineral Studies. *Vib. Spectrosc.* **2003**, *31*, 1–10. [CrossRef]

40. Thommes, M.; Kaneko, K.; Neimark, A.V.; Olivier, J.P.; Rodriguez-Reinoso, F.; Rouquerol, J.; Sing, K.S. Physisorption of gases, with special reference to the evaluation of surface area and pore size distribution (IUPAC Technical Report). *Pure Appl. Chem.* **2015**, *87*, 1051–1069. [CrossRef]

41. Parolo, M.E.; Avena, M.J.; Savini, M.C.; Baschini, M.T.; Nicotra, V. Adsorption and circular dichroism of tetracycline on sodium and calcium-montmorillonites. *Colloid Surf. A* **2013**, *417*, 57–64. [CrossRef]

42. Chang, P.H.; Li, Z.; Jiang, W.T.; Jean, J.S. Adsorption and intercalation of tetracycline by swelling clay minerals. *Appl. Clay Sci.* **2009**, *46*, 27–36. [CrossRef]

43. Parolo, M.E.; Savini, M.C.; Vallés, J.M.; Baschini, M.T.; Avena, M.J. Tetracycline adsorption on montmorillonite: pH and ionic strength effects. *Appl. Clay Sci.* **2008**, *40*, 179–186. [CrossRef]

44. Marco-Brown, J.L.; Barbosa-Lema, C.M.; Torres Sanchez, R.M.; Mercader, R.C.; dos Santos Afonso, M. Adsorption of picloram herbicide on iron oxide pillared montmorillonite. *Appl. Clay Sci.* **2012**, *58*, 25–33. [CrossRef]

45. Azizian, S. Kinetic models of sorption: A theoretical analysis. *J. Colloid Interface Sci.* **2004**, *276*, 47–52. [CrossRef] [PubMed]

46. Giles, C.H.; Smith, D.; Huitson, A. A general treatment and classification of the solute adsorption isotherm. I. Theoretical. *J. Colloid Interf. Sci.* **1974**, *47*, 755–765. [CrossRef]

47. Limousin, G.; Gaudet, J.P.; Charlet, L.; Szenknect, S.; Barthes, V.; Krimissa, M. Sorption isotherms: A review on physical bases, modeling and measurement. *Appl. Geochem.* **2007**, *22*, 249–275. [CrossRef]

© 2018 by the authors. Licensee MDPI, Basel, Switzerland. This article is an open access article distributed under the terms and conditions of the Creative Commons Attribution (CC BY) license (http://creativecommons.org/licenses/by/4.0/).

applied
sciences

MDPI

Article

Separation of Light Liquid Paraffin C₅–C₉ with Cuban Volcanic Glass Previously Used in Copper Elimination from Water Solutions

Miguel Autie-Pérez [1,*], Antonia Infantes-Molina [2], Juan Antonio Cecilia [2], Juan M. Labadie-Suárez [1] and Enrique Rodríguez-Castellón [1,*]

[1] Departamento FQB, Facultad de Ingeniería Química, Instituto Superior Politécnico José Antonio Hechevarría, MES, La Habana 19390, Cuba; Labadie.Bol2014@gmail.com
[2] Departamento de Química Inorgánica, Cristalografía y Mineralogía (Unidad Asociada al ICP-CSIC), Facultad de Ciencias, Universidad de Málaga, Málaga 29071, Spain; ainfantes@uma.es (A.I.-M.); jacecilia@uma.es (J.A.C.)
*   Correspondence: aautie@gmail.com (M.A.-P.); castellonl@uma.es (E.R.-C.); Tel.: +53-7-267-6104 (M.A.-P.); +34-952-131-873 (E.R.-C.)

Received: 10 January 2018; Accepted: 13 February 2018; Published: 17 February 2018

Featured Application: In this work, an inexpensive and available material, as volcanic glass, is used to absorb metals from wastewater and then it is used to the separation of light liquid-olefins.

Abstract: Raw porous volcanic glass from Cuba was used as an adsorbent for Cu^{2+} removal from dyes after activation with an acid solution. After Cu^{2+} adsorption, it was also evaluated its capacity to separate *n*-paraffins from a mixture by inverse gas chromatography (IGC), and the results were compared with those obtained with bare volcanic glass without copper. The main goal of this work is to highlight the great applicability of natural volcanic glass, which can be reused without pretreatment as an adsorbent. The results from copper adsorption were quite promising, considering the availability and low cost of this material; the sample without acid treatment turned out to be the most adequate to remove copper. Moreover, the results from IGC revealed that the separation of paraffins from the mixture was achieved with both bare volcanic glass and glass containing Cu, although greater heat adsorption values were obtained when copper was present in the sample due to the stronger interaction between paraffin and copper. The high availability and low cost of this porous material make it a potential and attractive candidate to be used in both heavy metal removal and paraffin separation for industrial purposes.

Keywords: glass; adsorption; surface properties; copper removal; IGC; paraffins

---

1. Introduction

Despite the decline in oil reserves, dependence on the petrochemical industry is still high, since there is no alternative energy source able to satisfy the demands of the world's population. Treatment of crude gives rise to a range of valuable products, such as fuels, waxes, asphalt, polymers, plastics, paints, tires, and detergents. Among the most important oil fractions are the short paraffins (C_nH_{2n+2}), since these compounds are main components of fuel and natural gas that have relatively low toxicity, because their combustion only generates CO_2.

A challenge for the petrochemical industry is olefin/paraffin separation, since both display similar physicochemical properties [1]. The use of membranes, cryogenic distillation, and absorption have emerged as potential methods for the separation of paraffin/olefin [2–5]. Among them, the most sustainable technology is related to selective adsorption of olefins, since its insaturation favors stronger interactions with the adsorbent [4]. However, the interaction of paraffins is more limited, since these compounds are only formed of C–C and C–H bonds, so the adsorption should take place through nonspecific interactions.

Traditionally, molecular sieves such as zeolites [6,7], activated carbons [8,9], and metal-organic frameworks (MOFs) [10,11], which display cavities with a specific pore size, have been used as efficient sieves to separate paraffin. However, these compounds also have some drawbacks related to the low thermal stability of the MOFs and the cost of synthesizing molecular sieves with appropriate pore size at a large scale. Currently, companies demand competitive adsorbents with lower production costs in order to increase the profit margin.

Natural clay minerals and zeolites are abundant and inexpensive natural molecular sieves that have been used in several adsorption processes [12–14] and paraffin separation [7]. Similarly, volcanic glass, a less well-known material than zeolites or clay minerals, has shown interesting behavior in the adsorption of harmful cations [15,16] and dyes [17–19] and in the separation of olefin/paraffin [20]. So volcanic glass should be considered as a potential adsorbent to separate *n*-paraffins. The genesis of volcanic glass takes place by a sudden cooling of the magma, obtaining a quasi-stable and amorphous structure [21]. The most abundant volcanic glass is obsidian, which was used as a knife blade or scalpel blade since prehistory [22,23]. Volcanic glass is a material that displays low thermal conductivity, low density, and high resistance to fire and perlite aggregate plasters. Hydration of obsidian leads to perlite, which shows a wider range of applications due to its high adsorption of sound, low bulk density, low thermal conductivity, and fire resistance, all of which provides interesting properties when used in the construction industry as plaster, insulation boards, and concrete, as well as a soil conditioner or carrier for herbicides, insecticides, and chemical fertilizers. In addition, perlite can be used for filtering water and other liquids, in food processing, and in pharmaceutical manufacturing [24]. Volcanic glass has also been used as starting material in the synthesis of clay minerals [25], as smectites, or as zeolites [26,27].

The aim of this research is to evaluate Cuban volcanic glass in two consecutives processes, the adsorption of transition metal cations and then its application to the separation of liquid paraffins. The adsorption capacity of volcanic glass was evaluated using Cu^{2+} cations as target metal. Cu species are considered pollutants from the mineral and metallurgic industries. The wastewater is a very dangerous environmental problem: it has nonbiodegradability features; it is hazardous to animal and human health, since the pollutants accumulate in living tissue, causing many harmful effects; it is implicated in biochemical and biological oxidation processes; and it causes serious problems in wastewater treatment plant sludge [28]. Several methods, such as precipitation, ion exchange, solvent extraction, and adsorption on oxides, have been shown to be highly efficient in heavy metal removal. Copper content should be less than 1.3 parts per million, since that is the limit for human consumption according to the US Environmental Protection Agency [29]. However, the high maintenance cost of these methods does not suit the need, mainly in developing countries [30].

Then, raw volcanic glass and Cu-volcanic glass were evaluated to separate liquid n-paraffins (C_5–C_9) by inverse gas chromatography (IGC) as alternative method to the traditional distillation employed in the petrochemical industry. The use of IGC to separate liquid paraffins has been studied in the literature, and interesting values have been attained for porous hexacyanocobaltates [31], metal-organic framework (Cu-BTC and Fe-BTC) [10], and even inexpensive materials such as natural clinoptilolite [7]. IGC evaluates both the surface and bulk properties of porous materials. This technique is highly efficient in analyzing the physicochemical properties of nonvolatile materials, such as diffusion coefficient, crystallinity, or specific parameters of surface free energy [32,33], which provides interesting information in nanomaterials [34,35], fibers [36], and polymer and coatings [37,38], as well as in the pharmaceutical industry.

2. Materials and Methods

2.1. Volcanic Glass

The studied material was volcanic glass extracted from Ají de la Caldera, Cuba. The volcanic glass, sieved to obtain a fraction lower than 0.1 mm, was activated with H_2SO_4 solution (0.2–0.6 mol L^{-1}) by stirring at 298 K for 4 h. Later, the samples were washed with water until the sulfate species were removed. Then the samples were dried in an oven at 383 K.

2.2. Characterization of the Materials

Elemental bulk composition of the sample was determined by wavelength-dispersive X-ray fluorescence (WDXRF) spectrometry using an ARL ADVANT'XP spectrometer under vacuum. Samples were dried and calcined and prepared as 35 mm diameter fused beads in a Katanax K1 machine. The composition was determined by using UniQuant software.

X-ray diffraction (XRD) patterns of the sample were obtained with an X'Pert PRO MPD diffractometer (PANanalytical) equipped with a Ge (111) primary monochromator using Cu $K_{\alpha 1}$ (1.5406 Å) radiation.

Diffuse Reflectance Infrared Fourier Transform (DRIFT) spectra were collected on a Harrick HVC-DRP cell fitted to a Varian 3100 FTIR spectrophotometer. The interferograms consisted of 120 scans, and the spectra were collected using a KBr spectrum as background. About 30 mg of finely ground sample was placed in a sample holder.

The textural properties (S_{BET}, V_p, d_p) of the volcanic glass were determined by N_2 adsorption-desorption at 77 K in a Micromeritics ASAP 2020 apparatus. Prior to analysis, volcanic glass was outgassed at 273 K and 0.01 kPa for 12 h. The specific surface area was determined using the Brunauer-Emmett-Teller equation considering an N_2 molecule cross-section of 16.2 $Å^2$ [39]. The total pore volume was calculated from the adsorption isotherm at $P/P_0 = 0.99$.

Size and morphology of the nanoparticles were studied by high-resolution transmission electron microscopy using a TALOS F200x instrument. TEM analysis was performed at 200 kV and 5.5 A and scanning transmission electron microscopy with a High Angle Annular Dark Field (HAADF) detector, at 200 kV and 200 nA.

X-ray photoelectron spectra were collected using a Physical Electronics PHI 5700 spectrometer with nonmonochromatic Mg $K\alpha$ radiation (300 W, 15 kV, and 1253.6 eV) with a multichannel detector. Spectra were recorded in the constant pass energy mode at 29.35 eV, using a 720 μm diameter analysis area. Charge referencing was measured against adventitious carbon (C 1s at 284.8 eV). A PHI ACCESS ESCA-V6.0 F software package was used for acquisition and data analysis.

2.3. Cu^{2+} Adsorption Tests

Cu^{2+} adsorption was carried out from a solution of $CuCl_2$ $2H_2O$ (1 mol L^{-1}) (Merck, analysis grade), using a range of concentrations between $1\ 10^{-3}$ and $1\ 10^{-1}$ mol L^{-1}.

The batch adsorption experiments were performed by immersing 100 mg of adsorbent with 10 mL of copper solution and stirring for 4 h. Then the solid was removed by filtration, while the liquid was analyzed in a UV-vis spectrophotometer, model Shimadzu UV mini-1240, at a wavelength of 600 nm. Cu adsorbed per unit mass adsorbent (mol g^{-1}) at equilibrium was calculated using the mass balance described by Equation (1):

$$q = \frac{V \times (c_0 - c_{eq})}{m_{ads}} \tag{1}$$

where V (mL) is the volume of the sample solution, c_0 and c_{eq} (mmol mL^{-1}) are the copper concentrations initially and after adsorption equilibrium is reached, q (mmol g^{-1}) is the amount of Cu adsorbed onto the volcanic glass, and m_{ads} (g) is the sample mass.

The adsorption isotherms were adjusted to the Langmuir model [40], assuming that each site can hold at most one molecule on a homogeneous surface, from Equation (2):

$$q = \frac{q_{max} \times K_L \times c_{eq}}{1 + (K_L \times c_{eq})} \tag{2}$$

where q_{max} (mmol g^{-1}) is the maximum adsorbed concentration, c_{eq} (mmol mL^{-1}) is the concentration of copper in solution at equilibrium, and K_L (L mg^{-1}) is the Langmuir constant.

2.4. Separation of N-Paraffin Tests

The IGC data were recorded with Shimadzu equipment (model 14B) and a flame ionization detector in the 403–488 K temperature range. Helium was used as a carrier gas at a flow rate of 12.8 cm^3/min. Previously cleaned and weighed stainless steel columns (60 cm long, 2.2 mm inner diameter) were packed with 2.5 g of the material. The packed columns were outgassed overnight at 523 K under helium flow. N-Alkanes of analytical grade were used in all cases and injected as the smallest detectable amount of their vapor phase extracted from the head space of their containers, to obtain data close to Henry adsorption coverage.

To obtain the corrected retention times for the probes, the relation $(t_R - t_0)$ was used, where t_0 is the retention time of methane, CH_4 (unretained hydrocarbon), and t_R is the retention time of the probe. The corrected retention times were taken as the average value among three injections. Mixtures of C_1-(C_5-C_9) were prepared by introducing a small quantity of CH_4 in the head space of each liquid alkane, and C_5-C_9 net retention volumes, V_n, were calculated according to Equation (3):

$$V_n = J \cdot V_f \cdot (t_R - t_0) \cdot \frac{T_c}{T_A} \cdot \frac{(P_0 - P_w)}{P_0} \tag{3}$$

where J is the James-Martin gas compressibility correction factor, V_f is the gas carrier flow rate at the flow meter temperature T_f, T_C is the column temperature, T_A is the room temperature ($T_A = T_f$), P_0 is the outlet pressure, and P_w is the vapor pressure of water at T_f.

The differential adsorption heat, Q_d, equal to the enthalpy for the standard adsorbed state ΔH_{d0} (within the Henry zone), was determined from the slope of $\ln (V_n/A_S)$ versus $1/T_C$ plot according to Equation (4):

$$Q_d = \Delta H_{d0} = -R \cdot \frac{d[\ln(V_n/A_S)]}{d(1/T_c)} \tag{4}$$

where R is the universal gas constant and A_S is the product of the specific surface area (S) and the amount of sample (m) in the column.

3. Results and Discussion

Characterization of Volcanic Glass

The elemental bulk composition of volcanic glass evaluated from X-ray Fluorescence Spectrometer (XRFS) analysis shows how this material is mainly composed and Si and Al, with small proportions of K, Fe, Ni, Ca, and Na. Previous research has pointed out that the viscosity of volcanic glass is directly related to the SiO_2 content [21]. Thus, volcanic glass with low SiO_2 content (<50%) is less viscous than glass with SiO_2 content above 65%. Considering the SiO_2 concentration shown in Table 1, the Cuban volcanic glass must display low fluency and high viscosity, probably due to slow cooling [21].

Appl. Sci. **2018**, *8*, 295

Table 1. Chemical composition of volcanic glass and Cu-volcanic glass (%).

Sample	SiO$_2$	Al$_2$O$_3$	K$_2$O	Fe$_2$O$_3$	CuO	MgO	CaO	Na$_2$O
Volcanic Glass	66.2	13.4	2.4	2.8	–	1.3	2.5	1.7
Cu-Volcanic Glass	63.3	12.2	3.6	3.4	2.2	1.9	1.8	1.2

The textural parameters of volcanic glass were determined from N$_2$ adsorption-desorption at 77 K (Figure 1). According to the International Union of Pure and Applied Chemistry (IUPAC) classification, the adsorption isotherm can be adjusted to type II [41], which is typical of nonporous or macroporous materials. Nonetheless, the N$_2$ adsorbed at lower relative pressure indicates that this volcanic glass also displays slight microporosity. The hysteresis loop can be considered as type H3, which is given by nonrigid aggregates of platelike particles (e.g., certain clays), but also if the pore network consists of macropores that are not completely filled with pore condensate [41]. The specific surface area, estimated by the Brunauer-Emmett-Teller equation [39], determined a low specific surface area (51 m^2/g), with a pore volume of 0.083 cm^3/g. After Cu exchange, the specific surface area was lower (19 m^2/g). Considering the low specific surface area and the macroporosity, the N$_2$ adsorbed must be attributed to the fillings between adjacent particles.

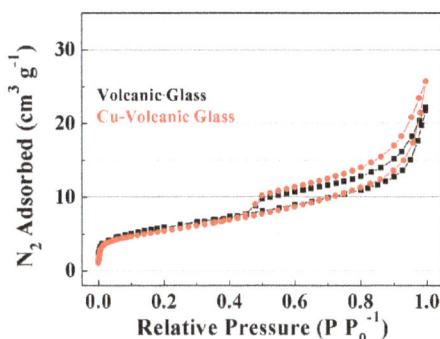

Figure 1. N$_2$ adsorption-desorption at 77 K of volcanic glass and Cu-volcanic glass.

According to the data of raw volcanic glass reported previously [20], XRD of volcanic glass hardly displays defined diffraction peaks. A broad band can be observed at 2θ = 20–30° which is attributed to the amorphous structure of volcanic glass. In addition, the existence of a pseudo-amorphous phase of aluminosilicate and Ca$_3$SiO$_5$ species, which shows higher crystallinity, is noteworthy.

As indicated previously [20], the FTIR spectrum of volcanic glass displays a broad signal between 3000 and 3700 cm^{-1}, which corresponds to –OH groups. Thus, the band located at 3620 cm^{-1} is assigned to structural OH-stretching, and those at 3420 cm^{-1} and 1640 cm^{-1} correspond to OH stretching and bending vibrational bands of adsorbed water [42]. After Cu incorporation, the broad signal remains, but the contribution located at 3620 cm^{-1} decreases in intensity, indicating that Cu interacts with OH groups. On the low wavelength side, Si–O stretching bands are located at 1148 cm^{-1}, 991 cm^{-1}, and 804 cm^{-1} [43,44]. The band at 908 cm^{-1} corresponds to the bending mode of Al–OH [43]. Raw volcanic glass also displays a weak shoulder band centered at ~1425 cm^{-1} and attributed to CO$_3{}^{-2}$ species, which must interact with the Ca^{2+} and Mg^{2+} species [45], reported in Table 1.

In order to evaluate the chemical composition of the surface of volcanic glass, X-ray Photoelectron Spectroscopy (XPS) analysis was performed for the raw material (Table 2). Similar to the elemental bulk composition (Table 1), Si and Al are the main elements in volcanic glass. In addition, the presence of other elements is noted, such as Mg and Ca, which can partially replace Al^{3+} sites by other Mg^{2+} ones, as well as carbonates and hydroxides, confirmed from the C 1s core level spectrum. The small amounts of Na⁺ and K⁺ can be used to counterbalance the charge deficiency of the aluminosilicate. In all cases, the binding energies are attributed to oxide species, except in the case of the O $1s$ core level spectrum, where two contributions, located at 532.4 and 531.4 eV, are detected (Figure 2) [46]. These contributions are attributed to the coexistence of oxide and hydroxide or carbonate species.

Table 2. Atomic concentration on the surface (%).

Sample	O $1s$	Si $2p$	Al $2p$	Mg $2p$	Ca $2p$	Na $1s$	K $1s$	Cu $2p$
Volcanic Glass	61.5	19.8	3.7	2.8	1.6	0.7	0.7	–
Cu-Volcanic Glass	62.1	19.9	3.8	2.2	0.9	0.7	0.7	0.6

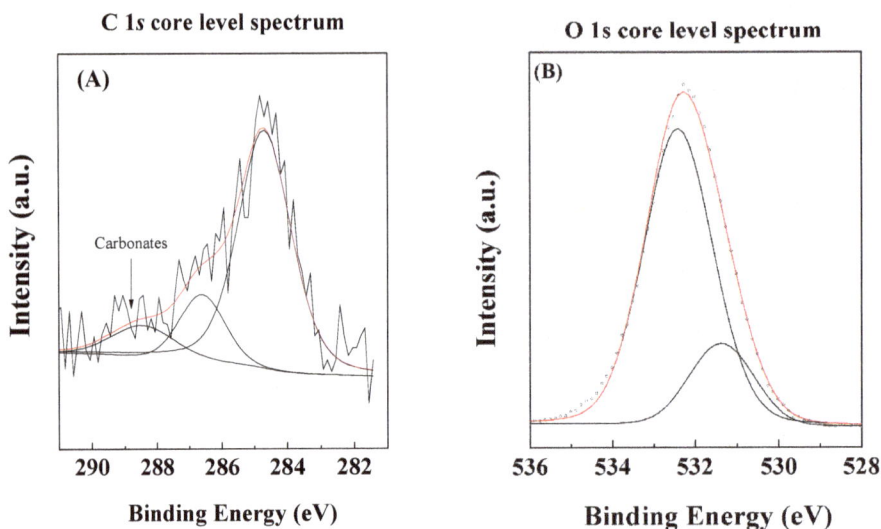

Figure 2. C $1s$ core level spectrum and O $1s$ core level spectrum of Cu-volcanic glass.

4. Cu²⁺ Adsorption Capacity of Volcanic Glass

The elemental composition of volcanic glass mainly consists of aluminosilicates, similar to clay minerals and zeolites, although its structure is amorphous. Previous research has established that an increase in temperature should cause an expansion of the amorphous structure; however, this thermal treatment can also produce dehydroxilation of the –OH groups, which are involved in the process of Cu adsorption [46], as indicated by the following reaction:

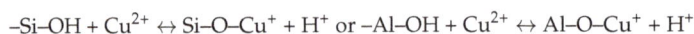

$$\text{–Si–OH} + \text{Cu}^{2+} \leftrightarrow \text{Si–O–Cu}^{+} + \text{H}^{+} \text{ or –Al–OH} + \text{Cu}^{2+} \leftrightarrow \text{Al–O–Cu}^{+} + \text{H}^{+}$$

The isotherm adsorption data (Figure 3) show how Cu adsorption decreases when volcanic glass is activated with more concentrated H_2SO_4 solutions, following the trend H_2SO_4 (0.0 M) > H_2SO_4 (0.2 M) > H_2SO_4 (0.4 M) > H_2SO_4 (0.6 M) due to modification of the point of zero charge of the volcanic glass [17,47]. These data are in agreement with the literature, since the heavy metal adsorption takes place in a narrow pH range [47–49]. In the case of adsorption of Cu²⁺ species, it has been reported that Cu adsorption increases at higher pH values [24,47]. Under these activation conditions, the leaching of volcanic glass must be discarded, since it requires stronger conditions [50,51].

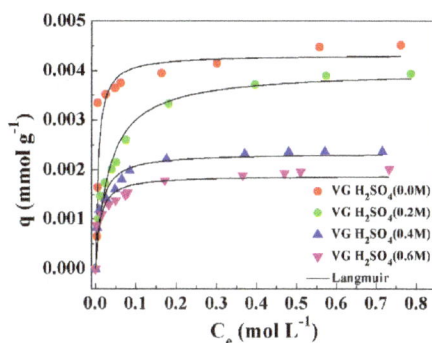

Figure 3. Cu^{2+} adsorption in volcanic glass (VG) with acid treatment and activated with H_2SO_4 (0.2 M, 0.4 M, and 0.6 M) solution.

The isotherm data were fitted to the Langmuir model (Table 3). These data reveal that volcanic glass without previous acid treatment reaches the q_{max} maximum value ($4.32 \cdot 10^{-4}$ mol g^{-1}); however, the q_{max} value decreases accordingly as the H_2SO_4 concentration used in the prior activation increases. The strength of the adsorbate-adsorbent interaction, defined by the K_L value, shows how volcanic glass without acid activation exhibits a stronger interaction.

Table 3. Equilibrium adsorption parameters according to the Langmuir model to Cu^{2+} adsorption on volcanic glass.

Sample	$q_{max} \times 10^3$ (mmol g^{-1})	K_L (L g^{-1})	R
Volcanic Glass (H_2SO_4 0.0 M)	4.32 ± 0.27	2.5 ± 0.9	0.96
Volcanic Glass (H_2SO_4 0.2 M)	4.00 ± 0.14	1.5 ± 0.5	0.97
Volcanic Glass (H_2SO_4 0.4 M)	2.34 ± 0.07	1.2 ± 0.2	0.96
Volcanic Glass (H_2SO_4 0.6 M)	1.81 ± 0.08	0.5 ± 0.1	0.94

Since volcanic glass without previous acid activation reached maximum adsorption capacity, this material was recycled to evaluate its behavior in liquid paraffin separation. Prior to this study, volcanic glass was characterized after the adsorption process to show the Cu adsorbed.

The elemental bulk composition of Cu-volcanic glass was evaluated from XRFS analysis (Table 1). These data indicate that the amount of Cu (in the form of CuO) after the adsorption process was 2.2%, while the amounts of other elements were close to that shown for raw volcanic glass, which discards the leaching of these elements along the adsorption process or along the acid activation. In the same way, the chemical composition on its surface was determined by XPS, reaching a Cu content of 0.6% (Table 2). The Cu 2*p* core level spectrum displays a main peak at about 932.8 eV, which is ascribed to Cu^{2+} species in the form of CuO, while the other contribution, located at about 935.2 eV, is assigned to the shake-up satellite, which is typical of bivalent cations [52]. From the atomic concentrations, a slight decrease of the Mg and Ca species on its surface can be observed, which could indicate partial leaching of these species or deposition of the Cu species in these zones.

Finally, the Cu content on its surface was also confirmed by TEM (Figure 4). This micrograph shows how Cu^{2+} species are distributed on the surface in some parts of volcanic glass; in other parts, it is not present. Energy-Dispersive X-ray spectroscopy (EDX) analysis confirms that the zones where copper is present correspond with the presence of carbon, and therefore its distribution on the surface depends on carbonate location. These data corroborate that Cu incorporation takes place from the intercalation of Cu with Ca from calcite in the pristine mineral. The data could be in agreement with the adsorption data, since the incorporation of Cu^{2+} species must take place by a cationic exchange of Mg^{2+} or Ca^{2+} by Cu^{2+} species, which is in agreement with the atomic concentrations on the surface (Table 2). The reaction that takes place with the carbonate species should be:

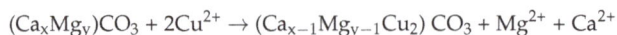

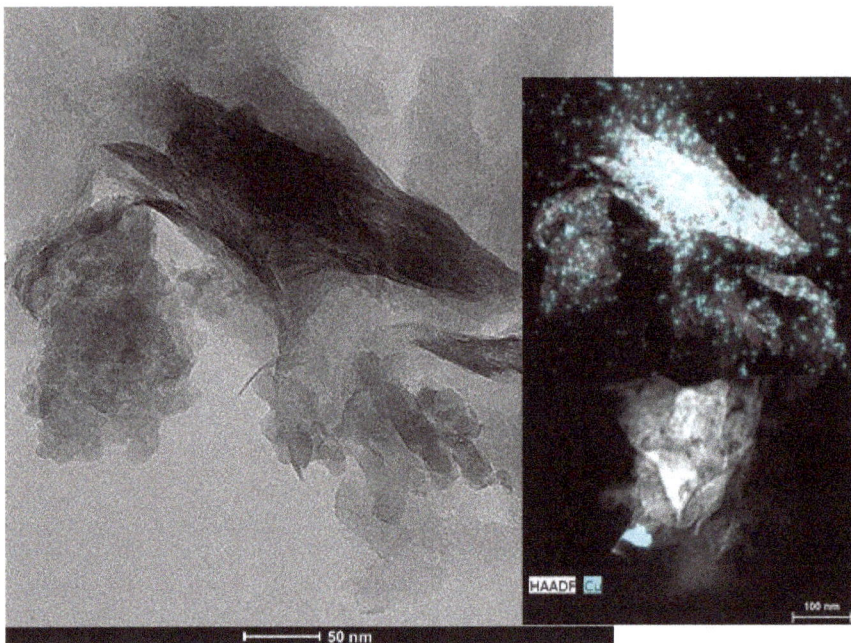

$$(Ca_xMg_y)CO_3 + 2Cu^{2+} \rightarrow (Ca_{x-1}Mg_{y-1}Cu_2) CO_3 + Mg^{2+} + Ca^{2+}$$

Figure 4. TEM micrograph of Cu-volcanic glass.

The adsorption isotherm of the raw volcanic glass reveals that Cu^{2+} species are adsorbed at low c_{eq} values (Figure 3). These data suggest a strong interaction between raw volcanic glass and the Cu^{2+} species, as was reported from the K_L values (Table 3). The acid activation of volcanic glass causes a decrease in the adsorption capacity, probably due to a slight dissolution of the carbonate species by the acid solution. In these cases, Cu^{2+} adsorption could be ascribed mainly to the interaction of the Cu^{2+} species with the silanol groups. The strength of this interaction is lower, as indicated by the isotherm profiles and the K_L values. However, K_L values for this material are in line with those found in the literature for other materials: 0.024 L/mg for chitosan-based materials [53]; 8–10 L/mg, depending on the temperature, for functional Poly-Ethylene Terephthalate (PET) fibers [54]; 0.002–0.0006 L/mg for carbon nanotube–mullite composites [55]; 0.015–0.239 L/mg for multiadsorbent systems containing tea waste and dolomite [56]; and 0.098–0.26 for nanocomposites of ZnO with montmorillonite [57].

To sum up, volcanic glass is an efficient material for the adsorption of heavy metals such as Cu. The presence of minor impurities of carbonate species improves Cu adsorption capacity by stronger chemical interactions. These Cu adsorption values are below those of other adsorbents, such as zeolites and activated carbons, due to these materials display high specific surface area and/or microporosity. Nonetheless, volcanic glass displays quite interesting Cu adsorption values, taking into account that this material is inexpensive and highly available.

5. Light N-Paraffin Separation in Volcanic Glass

Cuban volcanic glass and volcanic glass with higher Cu adsorption capacity, i.e., the sample without acid treatment activation, were chosen to evaluate their capacity to separate *n*-paraffins. The chromatographic profiles of the *n*-paraffin mixtures are compiled in Figure 5 for raw volcanic glass and Cu-volcanic glass. Taking into account the profiles of all experiments, it can be observed how each paraffin is eluted at a different retention time, so both samples are appropriate materials for the separation of n-paraffins. In all cases, the chromatographic peaks were sufficiently sharp and symmetric to consider the Henry zone, since the interactions with neighboring molecules must be nonspecific or show low intensity.

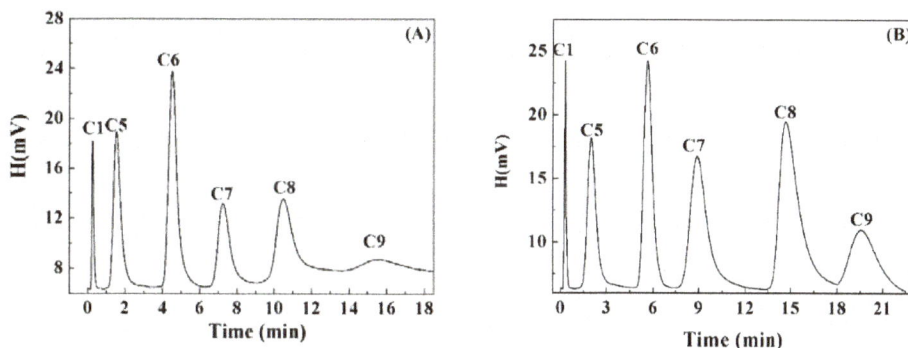

Figure 5. Chromatograms of CH_4 + C_5–C_9 mixture in (**A**) volcanic glass and (**B**) Cu^{2+}-volcanic glass. T_C = 400 K + 5 K/min.

As indicated, the adsorption of *n*-paraffins, which are only composed of C–C and C–H bonds, must be governed by nonspecific interactions. On the other hand, volcanic glass is an amorphous structure, mainly macroporous, although it also displays a certain microporosity (Figure 1). Thus, it is expected that the microporosity of volcanic glass can act as a molecular sieve for the separation of *n*-paraffins, as occurs with activated carbon and zeolites. In addition, it must be considered that the structure of volcanic glass expands with increased temperature, which could facilitate accessing n-paraffins inside the pores of volcanic glass. This can improve diffusion of the paraffins along the volcanic glass column, diminishing retention time.

The representation of (ln Vn) vs. (1/Tc) for each n-paraffin (Figure 6) can be fitted linearly, with R > 0.99 in all cases. The differential adsorption heat was determined from the slope, as shown in Table 3. These data reveal that adsorption heat is directly related to the length of the hydrocarbon, with the highest values obtained for nonane.

$$Q_d\ C_9H_{20} > Q_d\ C_8H_{18} > Q_d\ C_7H_{16} > Q_d\ C_6H_{14} > Q_d\ C_5H_{12}$$

Table 3 also reports how Cu-volcanic glass reaches higher adsorption heat values compared with raw volcanic glass. It is well known that transition metals, such as Ag, Cu, Co, and Ni, tend to interact with the double bond of the olefins [58]. The presence of these transition metals favors selective adsorption between olefins and saturated hydrocarbons with similar physicochemical properties, as takes place in propane-propylene separation. In the case of paraffins, the interactions between the Cu^{2+} species and C–H and C–H bonds should be negligible. Recently, Moreno-González et al. reported the existence of Cu^{2+}-alkane interactions by hyperfine sublevel correlation spectroscopy and electron paramagnetic resonance. Thus, it would be expected that the interaction between liquid *n*-paraffins and Cu^{2+} would lead to greater retention time and increased adsorption heat [59].

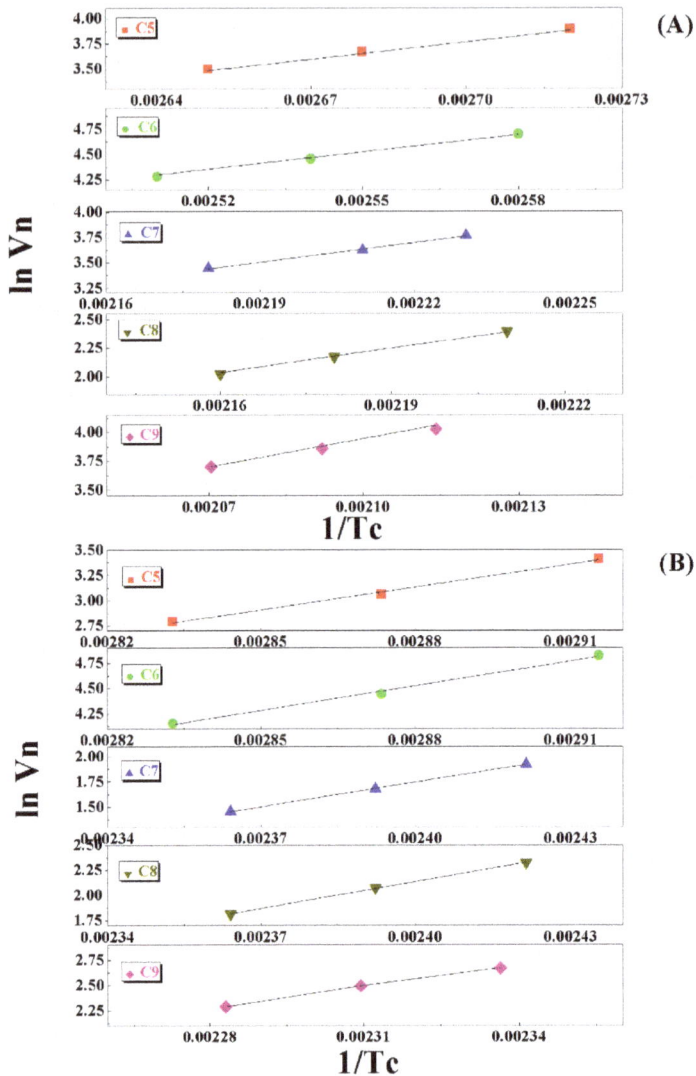

Figure 6. Ln Vn vs. 1/Tc representation by C_5–C_9 in (**A**) volcanic glass and (**B**) Cu^{2+}-volcanic glass.

The contribution of each $-CH_2-$ of the *n*-paraffins can be determined from the slope of the representation (adsorption heat vs. type of n-paraffin) (Figure 7), obtaining a Q_{dif} of 5.3 kJ/mol for raw volcanic glass, while Cu-volcanic glass only reaches a Q_{dif} of 4.2 kJ/mol (Table 4). In this way, previous research separated propane–propylene by inverse gas chromatography, indicating that propylene, which displays a stronger adsorbent–adsorbate interaction by the presence of the double bond, exhibits lower Q_{dif} in comparison to propane [20]. In this sense, the incorporation of Cu species into volcanic glass also causes a stronger interaction with n-paraffins, which could be related to a decrease of its Q_{dif} value. The variations of adsorption heat values between the n-paraffins are much higher than 1, confirming that both samples display an excellent capacity for separating *n*-paraffins.

Table 4. Fitting parameters of the adsorption isotherms of C_5–C_9 on volcanic glass and Cu^{2+}-volcanic glass at 273 K.

Sample	N-Paraffin	y = ax + b		r	Q_{dif}
		a	b		
Volcanic Glass	C5	5560	−11.23	0.995	46.2
	C6	6170	−11.23	0.997	51.3
	C7	6670	−11.11	0.995	55.4
	C8	7406	−13.97	0.993	61.5
	C9	8117	−13.10	0.995	67.4
Cu^{2+}-Volcanic Glass	C5	6932	−16.83	0.991	57.6
	C6	7410	−16.81	0.990	61.6
	C7	7810	−16.98	0.998	64.9
	C8	8669	−17.93	0.998	69.5
	C9	8969	−17.41	0.998	74.5

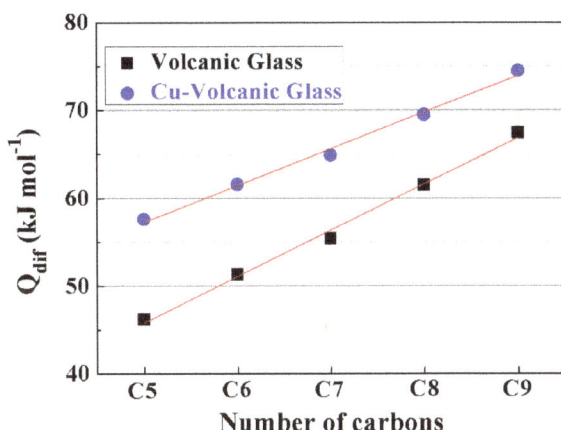

Figure 7. Representation of Q_{dif} vs. carbon number of the paraffin (C_5–C_9).

The adsorption heat of raw volcanic glass and Cu-volcanic glass was compared with other data reported in the literature using inverse gas chromatography under similar reaction conditions. Thus, the adsorption heat of both samples is in the same range as MOFs such as Cu–BTC [10] and natural zeolites such as clinoptilolite [7], so the use of inverse gas chromatography with inexpensive materials such as volcanic glass is a sustainable alternative to the traditional distillation to separate olefins/paraffins or several n-paraffins with different alkyl chains [20].

6. Conclusions

In the present paper, the use of natural volcanic glass from Cuba as an adsorbent in the removal of Cu^{2+} from dyes and its subsequent use to separate a mixture of *n*-paraffins by IGC were evaluated. It achieved great results in the copper removal experiments, related to the calcium carbonate content present. Copper replaces calcium from calcite during the adsorption process. Moreover, due to the partial solution of carbonates after acid activation, the sample without acid pretreatment achieved better copper removal results. Moreover, the Cu-containing sample was further studied in IGC to separate olefins, and great separation results were achieved. The results were also compared with bare mineral. In both cases, separation of the mixture was achieved. Several factors play important roles in the separation process: microporosity, hydroxy groups, and copper species. Its high availability, low cost, capacity to operate at low temperatures, and stability make this material a potential and attractive candidate to be used in both heavy metal removal and paraffin separation for industrial purposes.

Acknowledgments: Thanks to project CTQ2015-68951-C3-3R (Ministerio de Economía y Competitividad, Spain, and FEDER Funds). A.I.M. thanks the Ministry of Economy and Competitiveness for a Ramón y Cajal contract (RyC2015-17870).

Author Contributions: M.A. and J.M.L.-S. designed the experiments and prepared the materials. J.A.C., A.I.-M., and E.R.-C. performed the experiments and characterizations and wrote the paper in collaboration with M.A.

Conflicts of Interest: The authors declare no conflict of interest.

References

1. Bryan, P.F. Removal of Propylene from Fuel-Grade Propane. *Sep. Purif. Rev.* **2004**, *33*, 157–182. [CrossRef]
2. Merkel, T.C.; Blanc, R.; Ciobanu, I.; Firat, B.; Suwarlim, A.; Zeid, J. Silver salt facilitated transport membranes for olefin/paraffin separations: Carrier instability and a novel regeneration method. *J. Membr. Sci.* **2013**, *447*, 177–189. [CrossRef]
3. Eldridge, R.B. Olefin/paraffin separation technology: A review. *Ind. Eng. Chem. Res.* **1993**, *32*, 2208–2212. [CrossRef]
4. Azhin, M.; Kaghazchi, T.; Rahmani, M. A review on olefin/paraffin separation using reversible chemical complexation technology. *J. Ind. Eng. Chem.* **2008**, *14*, 622–638. [CrossRef]
5. Faiz, R.; Li, K. Olefin/paraffin separation using membrane based facilitated transport/chemical absorption techniques. *Chem. Eng. Sci.* **2012**, *73*, 261–284. [CrossRef]
6. Ferreira, A.F.P.; Mittelmeijer-Hazeleger, M.C.; Granato, M.A.; Martins, V.F.D.; Rodrigues, A.E.; Rothenberg, G. Sieving di-branched from mono-branched and linear alkanes using ZIF-8: Experimental proof and theoretical explanation. *Phys. Chem. Chem. Phys.* **2013**, *15*, 8795–8804. [CrossRef] [PubMed]
7. Rivera, A.; Farías, T.; de Ménorval, L.C.; Autié-Castro, G.; Yee-Madeira, H.; Contreras, J.L.; Autié-Castro, M. Acid natural clinoptilolite: Structural properties against adsorption/separation of n-paraffins. *J. Colloid Interface Sci.* **2011**, *360*, 220–226. [CrossRef] [PubMed]
8. Huang, H.; He, Z.; Yuan, H.; Chen, Y.; Kobayashi, N. Evaluation of n-butane gas adsorption performance of composite adsorbents used for carbon canister. *Procedia Eng.* **2011**, *18*, 78–85. [CrossRef]
9. Yahia, S.B.; Ouederni, A. Hydrocarbons gas storage on activated carbons. *Int. J. Chem. Eng. Appl.* **2012**, *3*, 220–227. [CrossRef]
10. Herm, H.; Bloch, E.D.; Long, J.R. Hydrocarbon separation in metal-organic frameworks. *Chem. Mater.* **2014**, *26*, 323–338. [CrossRef]
11. Bhadra, B.N.; Jhung, S.H. Selective adsorption of *n*-alkanes from *n*-octane on metal-organic frameworks: Length selectivity. *ACS Appl. Mater. Interf.* **2016**, *8*, 6370–6777. [CrossRef]
12. Rivera, A.; Farías, T.; de Ménorval, L.C.; Ruiz-Salvador, A.R. Preliminary characterization of drug support systems based on natural clinoptilolite. *Microporous Mesoporous Mater.* **2003**, *61*, 249–259. [CrossRef]
13. Elizalde-González, M.P.; Pérez-Cruz, M.A. Interaction between organic vapors and clinoptilolite–mordenite rich tuffs in parent, decationized, and lead exchanged forms. *J. Colloid Interface Sci.* **2007**, *312*, 317–325. [CrossRef] [PubMed]
14. Hernández, M.A.; Corona, L.; González, A.I.; Rojas, F.; Lara, V.H.; Silva, F. Quantitative study of the adsorption of aromatic hydrocarbons (Benzene, Toluene, and *p*-Xylene) on dealuminated clinoptilolites. *Ind. Eng. Chem. Res.* **2005**, *44*, 2908–2916. [CrossRef]
15. Ruggieri, F.; Marín, V.; Gimeno, D.; Fernández-Turiel, J.L.; García-Vallés, M.; Gutiérrez, L. Application of zeolitic volcanic rocks for arsenic removal from water. *Eng. Geol.* **2008**, *101*, 245–250. [CrossRef]
16. Steinhauser, G.; Bichler, M. Adsorption of ions onto high silica volcanic glass. *Appl. Radiat. Isotopes* **2008**, *66*, 1–8. [CrossRef] [PubMed]
17. Dogan, M.; Alkan, M. Removal of methyl violet from aqueous solution by perlite. *J. Colloid Interface Sci.* **2003**, *267*, 32–41. [CrossRef]

18. Alkan, M.; Dogan, M. Adsorption kinetics of Victoria blue onto perlite. *Fresenius Environ. Bull.* **2003**, *12*, 418–425.

19. Dogan, M.; Alkan, M.; Turkylmaz, A.; Ozdemir, Y. Kinetics and mechanism of removal of methylene blue by adsorption onto perlite. *J. Hazard. Mater.* **2004**, *109*, 141–148. [PubMed]

20. Fernández-Hechevarría, H.M.; Labadie-Suárez, J.M.; Santamaría-González, J.; Infantes-Molina, A.; Autié-Castro, G.; Cavalcante, C.L., Jr.; Rodríguez-Castellón, E.; Autié-Castro, M. Adsorption and separation of propane and propylene by Cuban natural volcanic glass. *Mater. Chem. Phys.* **2015**, *168*, 132–137. [CrossRef]

21. Friedman, I.; Long, W. Volcanic glasses, their origins and alteration processes. *J. Non-Cryst. Solid* **1984**, *67*, 127–133. [CrossRef]

22. Buck, B.A. Ancient technology in contemporary surgery. *West. J. Med.* **1982**, *136*, 265–269. [PubMed]

23. Disa, J.J.; Vossoughi, J.; Goldberg, N.H. A comparison of obsidian and surgical steel scalpel wound healing in rats. *Plast. Reconstr. Surg.* **1993**, *92*, 884–887. [CrossRef] [PubMed]

24. Alkan, M.; Dogan, M. Adsorption of copper(II) onto perlite. *J. Colloid Interface Sci.* **2001**, *243*, 280–291. [CrossRef]

25. Tomita, K.; Yamane, H.; Kawano, M. Synthesis of smectite from volcanic glass at low temperature. *Clay Clay Miner.* **1993**, *41*, 655–661. [CrossRef]

26. Yoshida, A.; Inoue, K. Formation of faujasite-type zeolite from ground Shirasu volcanic glass. *Zeolites* **1986**, *6*, 467–473. [CrossRef]

27. Yoshida, A.; Inoue, K. Whiteness in zeolite A prepared from Shirasu volcanic glass. *Zeolites* **1988**, *8*, 94–100. [CrossRef]

28. Rivas, L.; Villegas, S.; Ruf, B.; Peric, I. Removal of metal ions with impact on the environment by wate-insolutble functional copolymer: Synthesis and metal ion uptake properties. *J. Chil. Chem. Soc.* **2007**, *52*, 1164–1168. [CrossRef]

29. Ground Water and Drinking Water. Available online: https://www.epa.gov/ground-water-and-drinking-water/national-primary-drinking-water-regulations (accessed on 16 February 2018).

30. Bereket, G.; Aroguz, A.Z.; Ozel, M.Z. Removal of Pb(II), Cd(II), Cu(II), and Zn(II) from aqueous solutions by adsorption on bentonite. *J. Colloid Interface Sci.* **1997**, *187*, 338–343. [CrossRef] [PubMed]

31. Autié-Castro, G.; Autié, M.; Reguera, E.; Santamaría-González, J.; Moreno-Tost, R.; Rodríguez-Castellón, E.; Jiménez-López, A. Adsorption and separation of light alkane hydrocarbons by porous hexacyanocobaltates (III). *Surf. Interf. Anal.* **2009**, *41*, 730–734. [CrossRef]

32. Voelkel, A.; Strzemiecka, B.; Adamska, K.; Milczewaska, K. Inverse gas chromatography as a source of physiochemical data. *J. Chromatogr. A* **2009**, *1216*, 1551–1566. [CrossRef] [PubMed]

33. Mohammadi-Jan, S.; Waters, K.E. Inverse gas chromatography applications: A review. *Adv. Colloid Interface Sci.* **2014**, *212*, 21–44. [CrossRef] [PubMed]

34. Batko, K.; Voelkel, A. Inverse gas chromatography as a tool for investigation of nanomaterials. *J. Colloid Interface Sci.* **2007**, *315*, 768–771. [CrossRef] [PubMed]

35. Menzel, R.; Lee, A.; Bismarck, A.; Shaffer, M.S.P. Inverse gas chromatography of as-received and modified carbon nanotubes. *Langmuir* **2009**, *25*, 8340–8348. [CrossRef] [PubMed]

36. Heng, J.Y.Y.; Pears, D.F.; Thielmann, F.; Lampke, T.; Bismark, A. Methods to determine surface energies of natural fibres: A review. *Compos. Interface* **2007**, *14*, 581–604. [CrossRef]

37. Murakami, Y.; Enoki, R.; Ogoma, Y.; Kondo, Y. Studies on interaction between silica gel and polymer blend by inverse gas chromatography. *Polym. J.* **1998**, *30*, 520–525. [CrossRef]

38. Abel, M.L.; Chehimi, M.M.; Fricker, F.; Delamar, M.; Brown, A.M.; Watts, J.F. Adsorption of poly(methyl methacrylate) and poly(vinyl chloride) blends onto polypyrrole: Study by X-ray photoelectron spectroscopy, time-of-flight static secondary ion mass spectrometry, and inverse gas chromatography. *J. Chromatogr. A* **2002**, *969*, 273–285. [CrossRef]

39. Brunauer, S.; Emmett, P.H.; Teller, E. Adsorption of gases in multimolecular layers. *J. Am. Chem. Soc.* **1938**, *60*, 309–319. [CrossRef]

40. Langmuir, I. The adsorption of gases on plane surfaces of glass, mica and platinum. *J. Am. Chem. Soc.* **1918**, *40*, 1361–1403. [CrossRef]

41. Thommes, M.; Kaneko, K.; Neimark, A.V.; Olivier, J.P.; Rodriguez-Reinoso, F.; Rouquerol, J.; Sing, K.S.W. Physisorption of gases, with special reference to the evaluation of surface area and pore size distribution (IUPAC Technical Report). *Pure Appl. Chem.* **2015**, *87*, 1051–1069. [CrossRef]

42. Madejová, J. FTIR techniques in clay mineral studies. *Vib. Spectrosc.* **2003**, *31*, 1–10. [CrossRef]
43. Farmer, V.C. (Ed.) *Infrared Spectra of Minerals*; Mineralogical Society: London, UK, 1974; p. 331.
44. Vilarrasa-García, E.; Cecilia, J.A.; Azevedo, D.C.S.; Cavalcante, C.L., Jr.; Rodríguez-Castellón, E. Evaluation of porous clay heterostructures modified with amine species as adsorbent for the CO_2 capture. *Microporous Mesoporous Mater.* **2017**, *249*, 25–33. [CrossRef]
45. Correia, L.M.; Campelo, N.S.; Novaes, D.S.; Cavalcante, C.L., Jr.; Cecilia, J.A.; Rodríguez-Castellón, E.; Vieira, R.S. Characterization and application of dolomite as catalytic precursor for canola and sunflower oils for biodiesel production. *Chem. Eng. J.* **2015**, *269*, 35–43. [CrossRef]
46. Moulder, J.F.; Stickle, W.F.; Sool, P.E.; Bomber, K.D. *Handbook of X-Ray Photoelectron Spectroscopy*; Pekin-Elmer: Eden Praire, NM, USA, 1992.
47. Kinniburgh, D.G. *Adsorption of Inorganic Solid–Liquid Interfaces*; Anderson, M.A., Rubin, A.J., Eds.; Ann Arbor Science Publisher: Ann Arbor, MI, USA, 1981; pp. 91–160.
48. Ghassabzadeh, H.; Torab-Mostaedi, M.; Mohaddespour, A.; Maragheh, M.G.; Ahmadi, S.J.; Zaheri, P. Characterizations of Co(II) and Pb(II) removal process from aqueous solutions using expanded perlite. *Desalinitation* **2010**, *261*, 73–79. [CrossRef]
49. Ghassabzadeh, H.; Mohadespour, A.; Torab-Mostaedic, M.; Zaherib, P.; Maraghehc, M.G.; Taheric, H. Adsorption of Ag, Cu and Hg from aqueous solutions using expanded perlite. *J. Hazard. Mater.* **2010**, *177*, 950–955. [CrossRef] [PubMed]
50. Franco, F.; Pozo, M.; Cecilia, J.A.; Benítez-Guerrero, M.; Pozo, E.; Martín Rubí, J.A. Microwave assisted acid treatment of sepiolite: The role of composition and "crystallinity". *Appl. Clay Sci.* **2014**, *102*, 15–27. [CrossRef]
51. Franco, F.; Pozo, M.; Cecilia, J.A.; Benítez-Guerrero, M.; Lorente, M. Effectiveness of microwave assisted acid treatment on dioctahedral and trioctahedral smectites. The influence of octahedral composition. *Appl. Clay Sci.* **2016**, *120*, 70–80. [CrossRef]
52. Cecilia, J.A.; Arango-Díaz, A.; Rico-Pérez, V.; Bueno-López, A.; Rodríguez-Castellón, E. The influence of promoters (Zr, La, Tb, Pr) on the catalytic performance of $CuO-CeO_2$ systems for the preferential oxidation of CO in the presence of CO_2 and H_2O. *Catal. Today* **2015**, *253*, 115–125. [CrossRef]
53. Niua, Y.; Li, K.; Ying, D.; Wang, J.; Jia, J. Novel recyclable adsorbent for the removal of copper(II) and lead(II) from aqueous solution. *Bioresour. Technol.* **2017**, *229*, 63–68. [CrossRef] [PubMed]
54. Niu, Y.; Ying, D.; Li, K.; Wang, Y.; Jia, J. Fast removal of copper ions from aqueous solution using an eco–friendly fibrous adsorbent. *Chemosphere* **2016**, *161*, 501–509. [CrossRef] [PubMed]
55. Tofighy, M.A.; Mohammadi, T. Copper ions removal from water using functionalized carbon nanotubes–mullite composite as adsorbent. *Mater. Res. Bull.* **2015**, *68*, 54–59. [CrossRef]
56. Albadarin, A.B.; Mo, J.; Glocheux, Y.; Allen, S.; Walker, G.; Mangwandi, C. Preliminary investigation of mixed adsorbents for the removal of copper and methylene blue from aqueous solutions. *Chem. Eng. J.* **2014**, *255*, 525–534. [CrossRef]
57. Abubakar, H.; Ahmad, M.; Hussein, M.Z.; Ibrahim, N.A.; Musa, A.; Saleh, T.A. Nanocomposite of ZnO with montmorillonite for removal of lead and copper ions from aqueous solutions. *Process Saf. Environ. Prot.* **2017**, *109*, 97–105.
58. Grande, C.A.; Araujo, J.D.P.; Cavenati, S.; Firpo, N.; Basaldella, E.; Rodrigues, A.E. New π-complexation adsorbents for propane-propylene separation. *Langmuir* **2004**, *20*, 5291–5297. [CrossRef] [PubMed]
59. Moreno-González, M.; Palomares, A.E.; Chiesa, M.; Boronat, M.; Giamello, E.; Blasco, T. Evidence of a Cu^{2+}–alkane interaction in Cu-zeolite catalysts crucial for the selective catalytic reduction of NO_x with hydrocarbons. *ACS Catal.* **2017**, *7*, 3501–3509. [CrossRef]

© 2018 by the authors. Licensee MDPI, Basel, Switzerland. This article is an open access article distributed under the terms and conditions of the Creative Commons Attribution (CC BY) license (http://creativecommons.org/licenses/by/4.0/).

applied
sciences

MDPI

Article

Influence of Synthesis Parameters in Obtaining KIT-6 Mesoporous Material

Fernando R. D. Fernandes, Francisco G. H. S. Pinto, Ewelanny L. F. Lima, Luiz D. Souza, Vinícius P. S. Caldeira and Anne G. D. Santos *

Laboratory of Catalysis, Environment and Materials, Department of Chemistry, State University of Rio Grande do Norte, Mossoró/RN 59610-090, Brazil; fe.rodrigo@hotmail.com (F.R.D.F); gustavo_sk13@hotmail.com (F.G.H.S.P.); ewelannylouyde@hotmail.com (E.L.F.L.); luizsouza@uern.br (L.D.S.); viniciuscaldeira@uern.br (V.P.S.C.)
* Correspondence: annegabriella@uern.br; Tel.: +55-084-3315-2241

Received: 16 April 2018; Accepted: 30 April 2018; Published: 5 May 2018

Abstract: In the present work, modifications were made in the typical synthesis of KIT-6 mesoporous material. The molar ratio of P123 and its dissolution time, the type of alcohol, the aging time, and the heat treatment time were varied. The materials obtained were characterized by X-ray diffraction (XRD), thermogravimetry and differential thermogravimetry (TG/DTG), Fourier-transform infrared spectroscopy (FTIR), adsorption and desorption of N_2 and transmission electron microscopy (TEM). It was observed that the modifications interfered directly in the ordered structure of materials, displaying materials with cubic structures *Ia3d* (KIT-6 mesoporous material) and hexagonal structures *P6mm*, (SBA-15 mesoporous material). The type of alcohol probably acts on generation of the micelle, influencing the formation of the porous system and ordered structure. The results obtained indicate that the cubic structure of KIT-6 can be formed with a reduction of 30% in the P123 concentration, and decreases of 2, 6 and 12 h in times of P123 dissolution, aging and heat treatment, respectively. The modifications carried out in the synthesis procedure have resulted in ir being possible to employ materials with different characteristics, such as mesoscopic ordering and textural properties, in applications in the areas of catalysis and adsorption.

Keywords: mesoporous silica; KIT-6; synthesis parameters; cubic structure

1. Introduction

Mesoporous materials have been gaining considerable attention within the field of catalysis, adsorption and ion exchange, due to their unique properties and diverse applicability [1]. The KIT-6 mesoporous material has been intensely investigated by academics and industries in the last few years [2]. It has this name for having been discovered in the Korea Institute of Science and Technology [3]. It has a large average pore diameter (4–12 nm), wall thickness of 4 to 6 nm, and a tridimensional symmetric cubic structure *Ia3d* [4,5]. It also has a large pore volume, a high surface area and a bi continuous interpenetrating network, with additional micropores [2,6,7]. The pores of the KIT-6 provide easy and direct access to the active sites, which facilitates the insertion or diffusion of species in its interior [8].

The synthesis of KIT-6 was reported by Kleitz et al. [9]. In a typical hydrothermal synthesis of this material, hydrochloric acid is used to keep the medium acid, the triblock copolymer P123 as organic structure template, tetraethylorthosilicate (TEOS) as a source of silica; by contrast with materials such as SBA-15, butanol is used as a co-solvent and co-template. Each reagent has an indispensable function in the synthesis of KIT-6. The template enables the creation of materials with pores of different sizes and defined morphologies, thus producing channels that permeate the material [10]. The source of silica has the function of being the base material that will form the walls of the porous structure. In a typical synthesis of KIT-6, tetraethylorthosilicate (TEOS) is used due to its low cost and toxicity. The use of alcohol has the function of influencing the behavior of micelles during the synthesis process [11], facilitating P123 dissolution and helping in the organization of the cubic structure. The typical synthesis of KIT-6 follows pre-established times, which are necessary for the formation of its structure. Although there are no reports of works about the modification of reagents and stages of the KIT-6 synthesis process, there are several works about modifications in these parameters for similar mesoporous materials [12–15]. However, studies aimed at reducing the synthesis time for this material are relevant because, as seen by Impéror-Clerc et al. [12] in a study about SBA-15, the micelles of the template (P123) are already present in the first minutes of the synthesis.

Various studies have been developed with materials similar to KIT-6 with the goal of modifying parameters and reagents in order to evaluate the behavior of such variations in the final material. In a study performed by Cao et al. (2009) [15], the authors substituted P123 in the synthesis of SBA-15 for copolymers of polyethylene oxide (EO) and methyl polyacrylate (MA), which showed good results, promoting a material with hexagonal structure and good pore size distribution. Modifications in the source of the silica, as well as in the formation and organization stages of these materials, are also documented in the literature [14,16–18]. Studies demonstrating modifications in the synthesis parameters and obtaining materials with different properties are also reported for other types of materials, such as ordered mesoporous carbon (OMC) [19,20]. Such studies show the importance of analyzing the synthesis parameters and how they interfere in the final material.

Considering the promising characteristics of KIT-6 mesoporous material and due to the fact that there are no reports about modifications of its synthesis, the present study aims to vary parameters of KIT-6 synthesis in order to evaluate these changes in the structure and properties of the material.

2. Materials and Methods

2.1. Synthesis of Standard KIT-6

The standard KIT-6 sample was obtained following the typical method described by Kleitz et al. [9], using the following molar ratio: 1.000 TEOS; 0.017 P123; 1.83 HCl; 195 H_2O; 1.31 butanol. For 0.05 kg of synthesis gel, 1.23 g of triblock copolymer pluronic (P123, EO_{20}-PO_{70}-EO_{20}, Sigma Aldrich, St. Louis, MI, USA) was dissolved in 2.25 g of hydrochloric acid (37% Vetec) and 44 g of distilled water, this solution was subjected to vigorous stirring at 308 K for 6 h. After complete dissolution, 1.21 g butanol (99.4% Vetec) was added. The mixture remained under stirring at 308 K for 1 h. Thereafter, the tetraethylorthosilicate (TEOS, 98% Sigma Aldrich, St. Louis, MI, USA) was added to the homogeneous solution. Then, this mixture was kept under stirring at 308 K for 24 h, and in sequence was transferred to an autoclave and heated at 373 K for 24 h under static conditions. The solid product finally obtained was separated by vacuum filtration, washed with 2% HCl/ethanol solution, and dried overnight at room temperature. Subsequently, the material was calcined at 823 K for 6 h in synthetic air atmosphere to remove the organic template. The sample obtained through the standard synthesis was coded as KIT-6.

2.2. Modifications in the Standard Synthesis

The changes in the parameters of the synthesis of KIT-6 were carried out starting from the synthesis adopted as standard [9], altering only one parameter at a time, keeping the others unchanged.

The molar ratio of organic template P123 was modified by changing the standard ratio of 0.017 to 0.012 and 0.022. Samples obtained through this procedure were coded as K0.012 and K0.022, respectively. The dissolution time of the organic template was systematically modified, altering the initial time from 6 h to 8, 4 and 2 h, coded by KDT8, KDT4 and KDT2, respectively. Another modified parameter was the standard alcohol (butanol), which was replaced by the following alcohols: methanol (Synth 99.4%), ethanol (Synth 99.5%), 1-propanol (99.5% Vetec) and 1-pentanol (99% Sigma Aldrich). These samples were coded as KMt, KEt, KPr and KPe, respectively. The aging time of the synthesis gel was also modified, varying the standard time from 24 h to 6, 12, 18 and 30, with KAT6, KAT12, KAT18 and KAT30 samples, respectively. The last modified parameter was the time of heat treatment, changing the initial time from 24 to 12, 18 and 30 h, which were coded as KHT 12, KHT 18 and KHT 30, respectively.

Figure 1 shows a representative scheme with a summary of all modified parameters in the KIT-6 synthesis, and the respective sample codes for each parameter.

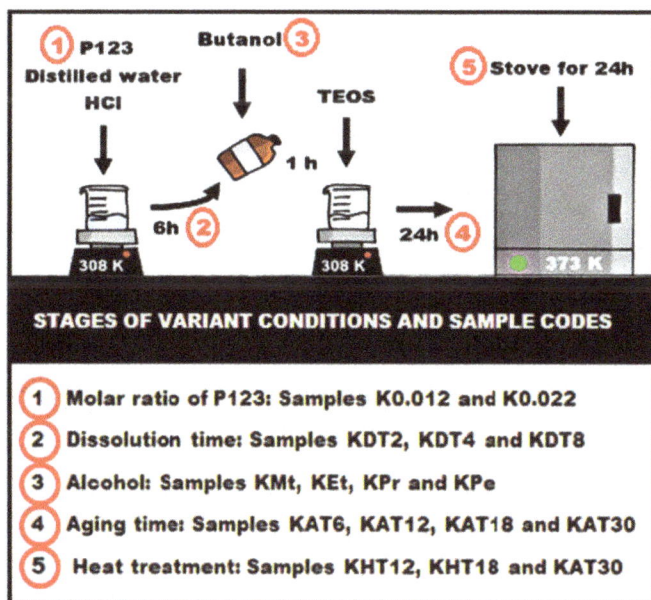

Figure 1. Schematic of the modifications made in the KIT-6 synthesis.

2.3. Characterizations

The X-ray diffraction (XRD) patterns were carried out on a Rigaku Mini Flex II equipment using CuKα radiation at a voltage of 30 KV and a 15 mA tube current. The data were collected at low-angle 0.5–3.0° 2θ, with step of 0.005° and time of 0.4 s. Thermogravimetric analysis (TGA) was performed on the STA 449F3-Jupiter equipment. About 5 mg of material was used in an alumina crucible under heating from 298–1173 K, with a heating rate of 10 K·min^{-1}, by a dynamic nitrogen atmosphere with flow of 25 mL·min^{-1}. The experiments of adsorption and desorption of N_2 were performed at 77 K on ASAP 2020 Micromeritics equipment. Prior to measurement, the samples were degassed at 573 K for 10 h under vacuum. The Brumauer–Emmett–Teller (BET) method was used to calculate the specific area [21] using the Rouquerol criterion to better apply this method [22]. The mesoporous volume was determined from the α-plot method for ordered mesoporous materials (α-plot OMM), and the total volume obtained by the Gurvich method [23]. The pore size distributions (PSD) were derived from the adsorption branch of the isotherms using the methods of Barrett–Joyner–Halanda (BJH), Villarroel–Bezerra–Sapag (VBS) [24] and the non-local density functional theory (NLDFT) [25]. For discussion of pore diameter, the VBS method was also applied in the desorption branch. Fourier-transform infrared spectroscopy (FTIR) spectra were obtained in the medium infrared region of 4000–400 cm^{-1} with 4 cm^{-1} resolution in a Bomem spectrophotometer model MB102. Transmission electron microscopy (TEM) was performed on Jeol microscope equipment, model JEM-2100 CM-200 (200 KV). Before analysis, the sample was dispersed in ethanol using low frequency ultrasound and placed in a copper screen coated with carbon.

Through the TEM images it was possible to determine the pore diameter for KDT4 and KHT12 samples. The values were measured from the ImageJ software, using the arithmetic mean for 50 points, with a standard deviation less than 20%.

3. Results and Discussions

Figure 2 shows the low-angle diffractograms of all the synthetized samples.

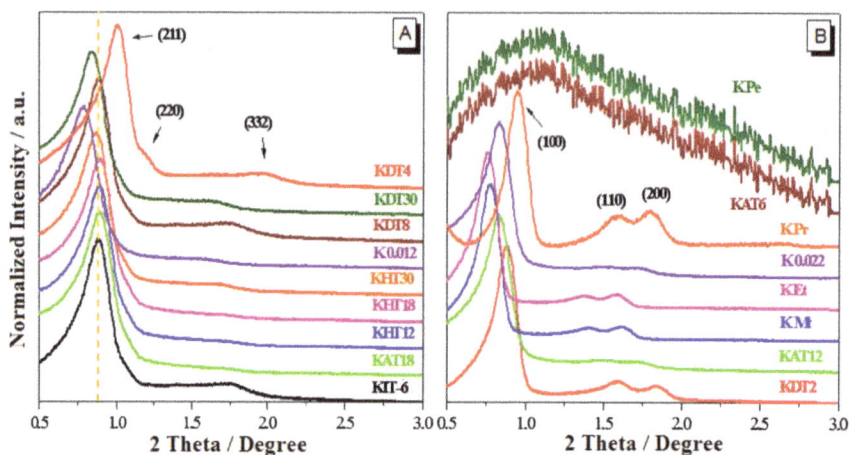

Figure 2. X-ray diffraction (XRD) patterns of all synthesized samples.

The samples in Figure 2A show reflection characteristics of materials with KIT-6 type, displaying (211), (220) and (332) Miller indices that are characteristic of three-dimensional mesoporous structures with spatial group *Ia3d*, typical of structures organized in cubic array [9,26,27]. The samples K0.022, KDT2, KMt, KEt, KPr and KAT12 in Figure 2B show reflections with (100), (110) and (120) Miller indices. According to Zhao et al. [28], these reflections are characteristic of two-dimensional mesoporous structures with spatial group *P6mm*, typical of structures organized in a hexagonal array, corresponding to SBA-15 type materials. The samples KAT6 and K Pe in Figure 2B not show any arranged structure.

Based on the data of the modification in the molar proportion of the organic template, two different behaviors can be noticed. The sample synthetized with a lower quantity of template (K0.012 sample) showed similar results to the standard KIT-6. This demonstrates that, even with the reduction employed in the synthesis, there was no interference in the critical micellar concentration (CMC) and subsequent formation of 3D cubic structure desired. A different behavior was observed with regard to the sample with a higher than standard molar proportion (K0.022 sample). The formation of a 2D hexagonal structure was noticed, referent to the SBA-15 mesoporous material. According to Wanka et al. [29], the increase in the quantity of the template used in the synthesis at constant temperature can lead to a transition in the cubic phase to the hexagonal phase, as can be observed in the present study. Figure 3A shows a representative scheme of the data obtained in this modification stage.

For the modification performed at the dissolution time of the organic template, it is noticeable that, by reducing the standard time from 6 h to 2 h, a cubic *Ia3d* structure is not obtained. However, for this reduction time a SBA-15 material with high *P6mm* hexagonal organization is obtained. This behavior can be related to the micellar formation and posterior protonation of the EO (ethylene oxide) groups in the early synthesis stage. The short period of 2 h, possibly, is not enough for the interaction of the micelles with the inorganic phase of silica, thus favoring the formation of a two-dimensional structure, which is thermodynamically more favorable than a three-dimensional one. For all the samples with times higher than 2 h, the formation of cubic structure-like KIT-6 material was observed (Figure 3B).

However, when observing the XRD peaks of the KDT4 sample, a displacement to a higher value of 2θ/angle in comparison to the KIT-6 standard was noted. This behavior demonstrates an interference in the structural organization of the material. Such results determine that the required time to the dissolution of the P123 template directly influences the formation and structural organization in the synthesis of the KIT-6 material.

The modifications made in the type of alcohol employed in the standard synthesis showed significant data regarding the study of the synthesis mechanism of ordered mesoporous material. It was possible to observe that all alcohols with carbonic chains shorter than butanol (KMt, KEt and KPr samples) exhibited a structure of SBA-15 type, while the alcohol with the higher carbonic chain, 1-pentanol (sample KPe), did not display any ordered structure. In a study developed by Kim et al. [30], the authors proposed that butanol interacts with the chains of PEO (polyethylene oxide) and PPO (polypropylene oxide) of the P123 micelles behaving as a type of co-template. Thus, the butanol alcohol acts directly on the curvature of the bi-continuous channels and is responsible for forming the cubic structure of the KIT-6 material. Considering this study and the results obtained in the present work, it has been shown that the type of alcohol is an important parameter for the generation of the *Ia3d* cubic structure. Due to the small size of the carbonic chains (methanol, ethanol and propanol), the interaction of the steric forces exerted by these alcohols in the micelle is possibly reduced. In this case, the smaller carbon chains are probably not enough to promote the posterior curvature of channels, resulting in unidirectional channels typical of SBA-15 materials. An opposite behavior was observed when increasing the size of the carbonic chain (KPe sample). The data probably suggest that pentanol alcohol provides an excessive interaction between the micelles to the point of promotion of the collapse of the structure. Figure 3C shows a representative scheme as a proposal of the influence of the variation on the type of alcohol in the formation mechanism of the KIT-6 material.

The modifications on the aging time of the synthesis gel showed that, with the increase of time, the formation of the cubic structure is favored. This behavior reveals that 6 h (KAT6 sample) is inadequate to the formation of any organized structure. This fact can be related to the low rate of hydrolysis and posterior condensation of the silica groups. From 12 h (KAT12 sample) of aging time the formation of a material with hexagonal phase and unidirectional channels like the SBA-15 material was noticed. This fact corroborates the results described by Kim et al. [30], where the authors observed the formation of a unidirectional phase with low synthesis time, which preceded the cubic phase. In addition to the present discussion, some studies showed that the change in synthesis parameters could cause a transition from the hexagonal to the cubic phase, and later to a lamellar phase [31,32]. Corroborating these studies, this study demonstrated that the X-ray diffraction behavior at longer aging times, such as 18 h and 30 h (KAT18 and KAT 30 samples), provided the formation of the *Ia3d* cubic structure characteristic of KIT-6 material. This case reveals that the "organic–inorganic" agglomerate obtained by the deposition of silicate species over the co-polymer needs periods longer than 18 h to go from a two-dimensional hexagonal organization with one-directional channels to a cubic three-dimensional phase with curved channels. Another relevant discussion refers to the curvature of the bi-continuous channels of the KIT-6 material. This indicates that the adequate time for there to be interaction forces exercised by the alcohol in the P123 micelles is between 12 h and 18 h of aging time in the synthesis gel. Considering this fact, even when using the standard alcohol for all the syntheses of this modification stage, only with times higher than 18 h can the desired *Ia3d* cubic structure be obtained (see scheme in Figure 3D).

The results of the variation on the thermal treatment time of synthesis gel show that all of the materials of this modification stage (KHT12, KHT18 and KHT30 samples) had structure typical of KIT-6, XRD peaks that were very similar to the standard sample. This fact indicates that all of the formation stages and structural organization of the cubic phase are probably found in the earlier stages of the synthesis mechanism. Thus, it is possible to obtain a high degree of condensation of the Si-OH groups and later formation of Si-O-Si bonds without any significant interference in the material's structure with a reduction of up to 12 h of the standard time. It should be noted that modifications performed in this parameter of the synthesis demonstrated greater influence on the textural properties (discussions performed posteriorly).

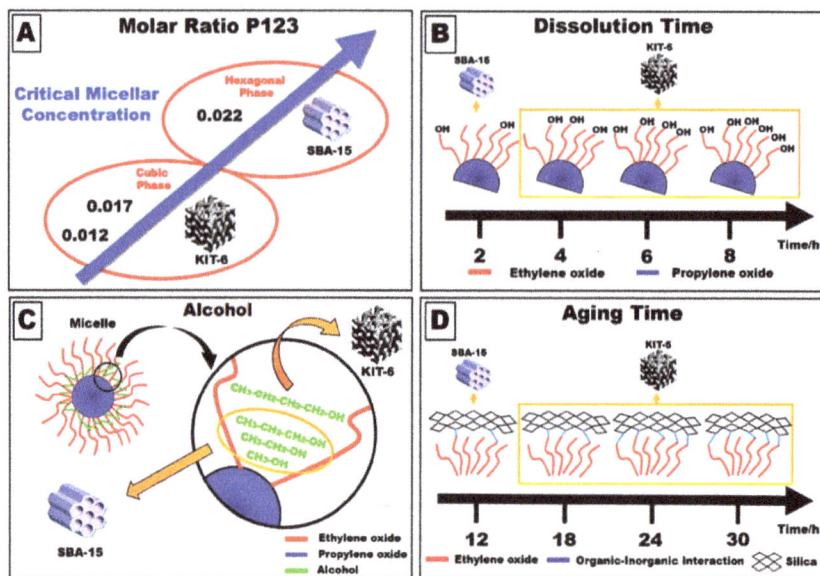

Figure 3. Representative scheme of the data obtained in: (**A**) the molar ratio of P123; (**B**) P123 dissolution time; (**C**) alcohol; and (**D**) aging time.

By the XRD data, the interplanar distance ($d_{(211)}$ and $d_{(100)}$), and the mesoporous parameters (a_{cub} and a_{hex}), for cubic and hexagonal structures, respectively, could also be determined. Table 1 exhibit these values for all the samples. In general, the samples showed cubic structures had very similar behavior to the standard KIT-6, except in some specific cases discussed below.

Table 1. Interplanar distance and mesoporous parameter of all samples with ordered structures.

Samples	$d_{(211)}$/nm	$d_{(100)}$/nm	a_{Cub}/nm	a_{Hex}/nm
KIT-6	5.14	-	12.60	-
K0.012	5.50	-	13.50	-
K0.022		10.70	-	12.37
KDT2		11.00	-	12.70
KDT4	4.53	-	11.10	-
KDT8	5.15	-	12.60	-
KMt		11.00	-	12.70
KEt		11.40	-	13.17
KPr		10.10	-	11.67
KAT12		10.80	-	12.30
KAT18	5.12	-	12.54	-
KAT30	5.20	-	12.92	-
KHT12	5.13	-	12.56	-
KHT18	5.14	-	12.60	-
KHT30	5.17	-	12.66	-

$d_{(211)}$ = interplanar distance calculated from $\lambda CuK\alpha = 2d_{(211)}$. sen θ; $d_{(100)}$ = interplanar distance calculated from $\lambda CuK\alpha = 2d_{(100)}$. sen θ; a_{Cub} = cubic parameter calculated from $a_{Cub} = d_{(211)}$. $6^{1/2}$; a_{Hex} = hexagonal parameter calculated from $a_{Hex} = 2d_{(100)}/3^{1/2}$.

The K0.012 and KAT30 samples presented interplanar distance and mesoporous parameter values higher than the standard KIT-6, while the sample KDT4 presented lower values. Relative to the decrease in P123 concentration (K0.012 sample), it can be determined that this variation directly interfered in the organization of the channels, increasing the distance between them. The data obtained for the KAT30 sample shows that the increase in the standard aging time interferes in the organization of the channels formed in the material. This fact can be related to the function of alcohol in the KIT-6 synthesis, which at long times can lead to an excessive curvature and consequently an influence in the structure, with 24 h being the ideal time. The behavior observed in the KDT4 sample denotes that the decrease in the time destined for the micellar formation directly modifies the organization of the future channels, generating significant reduction of 1.5 nm of the mesoporous parameter.

Even for the samples that had two-dimensional hexagonal structures, interesting behaviors can be observed. Analyzing the samples obtained from the alcohol modification, an increase on the values of $d_{(100)}$ and a_{hex} of the KMt compared to KEt samples can be observed, and a decrease on these values for KPr sample can be seen. This behavior demonstrates that the alcohol probably can act as a co-solvent or co-template, having a greater influence on the structure according to the size of its chain. Thus, alcohols with chains (C1–C2) that are too short do not interfere in a significant way on the hexagonal organization, indicating the behavior of a co-solvent. However, alcohols with longer chains interfere on the structural organization (the propanol (C3) in this case) or act in such a way that it completely modifies the structure of the material, except for the butanol (C4) that acts as a co-template.

The thermogravimetry and differential thermogravimetry (TG/DTG) curves and FTIR spectra for KIT-6, K0.012 and KDT4 representative samples has been represented in Figure 4A–C. The data of other samples can be seen in the supplementary material (Figures S1–S3).

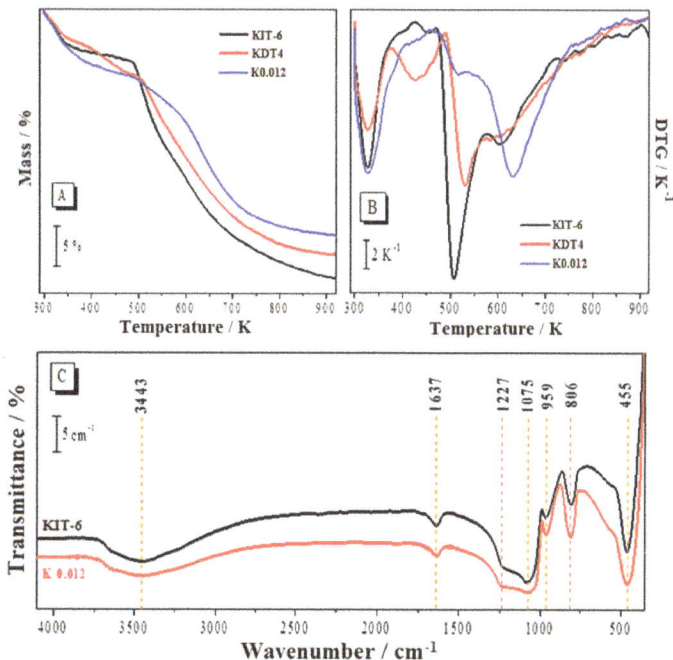

Figure 4. (**A**) Thermogravimetry (TG) and (**B**) differential thermogravimetry (DTG) curves of the KIT-6, K0.012 and KDT4 samples; and (**C**) the Fourier-transform infrared spectroscopy (FTIR) spectra of the standard KIT-6 and KDT2 samples.

In general, it can be observed that all samples had between two or more mass losses, according the Table 2.

Table 2. Quantification of mass loss events of all samples with ordered structures.

Samples	Events/K				Mass Loss/%			
	I	II	III	IV	I	II	III	IV
KIT-6	298–418	418–583	583–823	-	5.6	14.6	10.4	-
K0.012	298–443	443–538	538–823		7.7	2.6	16.6	
K0.022	298–383	383–546	546–823		12.2	5.0	15.3	
KDT2	298–383	383–565	565–823	-	4.4	24.0	14.6	-
KDT4	298–375	375–489	489–588	588–823	4.2	4.0	9.8	10.5
KDT8	298–460	460–600			4.6	14.6		
KAT12	298–388	388–598	598–823	-	10.0	17.6	4.0	-
KAT18	298–463	463–823	-	-	10.2	14.7	-	-
KAT30	298–500	418–610			4.6	16.6		
KMt	298–403	403–603	603–823		11.6	14.9	5.0	
KEt	298–398	398–583	583–823		9.8	18.0	4.3	
KPr	298–413	413–593	593–823		13.4	13.0	6.4	
KHT12	298–458	458–823	-	-	6.2	17.1	-	-
KHT18	298–458	458–823	-	-	9.0	17.4	-	-
KHT30	298–459	459–823	-	-	10.3	15.0	-	-

It can be observed that the standard KIT-6 showed three well-defined mass losses. The first loss corresponds to the removal of physisorbed water in the interior of the materials; the second loss is related to the degradation of the organic template and the removal of chemisorbed water; the third loss corresponds to the condensation of the remaining silanol groups in the surface and in the pores of the materials [33,34]. In general, the other samples showed similar losses, with some cases of specific changes in the temperature intervals of the mass losses and the quantity of losses being also observed. One of these cases was the KDT4 sample, in which four mass losses were observed. This higher difficulty on the removal of the organic template can be directly connected to the disorder found in its structure, which corroborates with the XRD results. Another pertinent discussion is in regard to the samples that formed hexagonal structures with unidirectional channels of the SBA-15 type, according to the XRD technique. All samples showed the second mass loss event at lower initial temperatures compared to the other samples with KIT-6 structure (see Figure S2 in the Supplementary Material). This fact can demonstrate a greater facility on the removal of the organic template due to the unidirectional system of its channels.

Figure 4C presents the FTIR spectra of the standard KIT-6 and the KDT4 and KDT2 representative samples for the ordered 3D cubic and unidirectional hexagonal structures. In general, bands related to the stretches of the inorganic functional groups, in both KIT-6 and SBA-15 materials, can be observed. In all samples were identified the absorption bands around 3443, 1637, 1227, 1075, 959, 806 and 455 cm^{-1}.

The band around 3443 cm^{-1} is attributed to the stretching vibrations of the hydroxyl groups, related to the O–H of the Si–OH groups found in the samples [35]. The band around 1637 cm^{-1} is attributed to the interaction of water with the support surface [36]. Bands around 1075 cm^{-1} and the identified shoulder at 1227 cm^{-1} are typical bands referring to asymmetric stretches of the Si–O–Si bonds. The band an 806 cm^{-1} are attributed to symmetrical stretches of the Si–O–Si bonds, and the bands near 959 and 455 cm^{-1} refer respectively to the symmetrical and asymmetric stretches of the Si–O bonds of the Si–OH groups [37].

The structural profile mentioned above was observed for both samples with 3D cubic structure and 2D hexagonal structure. This fact occurs because the materials have in their composition only bonds between Si and O atoms, including the formation of silanols (Si–OH) groups.

Figure 5 shows the results of adsorption and desorption analysis of N$_2$ at 77 K, performed on samples that presented ordered structures.

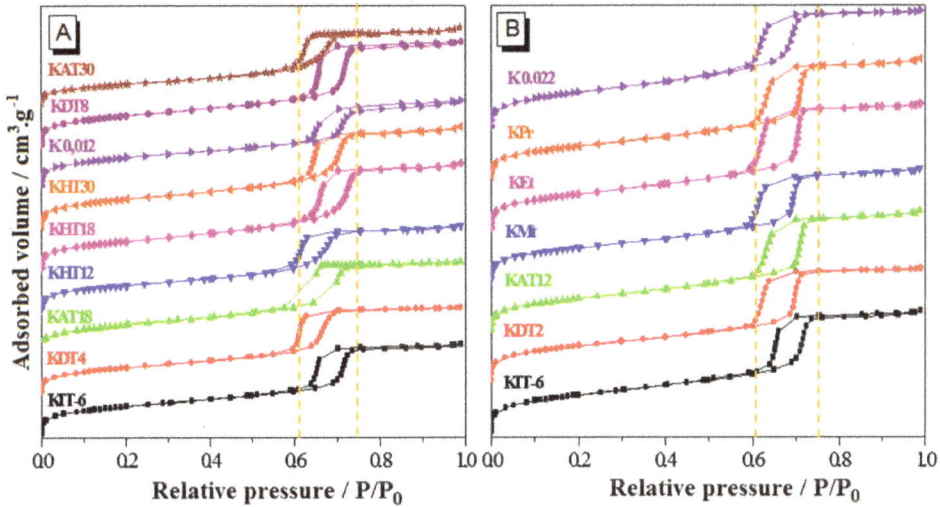

Figure 5. N_2 adsorption and desorption isotherms of the (**A**) samples that presented cubic structures and (**B**) samples that presented hexagonal structures.

From the isotherms, it can be observed that regardless of the modifications made during the synthesis procedure, all of the samples exhibited isotherms classified as type IV (a), characteristic of mesoporous materials with capillary condensation between 0.6 and 0.8 P/P_0, and a typical final saturation plateau [38]. As well, the referred capillary condensation was accompanied by H1 hysteresis loop. This hysteresis loop is characteristic of materials that show a cylindrical pores system [38]. Furthermore, it is relevant to highlight that all of the materials showed a very narrow hysteresis loops (between 0.6 and 0.75 P/P_0), which indicates a pores system with high uniformity. The exception was the KAT18 sample that exhibited a small loss of verticality in the hysteresis, which can be attributed to a porous system with higher distribution rate of average pore sizes.

Table 3 presents the textural properties of all samples with ordered structure.

Table 3. Textural properties of samples with ordered structures.

Samples	$S_{BET}/m^2 \cdot g^{-1}$	$V_{Micro}/cm^3 \cdot g^{-1}$	$V_{Meso}/cm^3 \cdot g^{-1}$	$V_T/cm^3 \cdot g^{-1}$
KIT-6	603	0.02	0.68	0.76
K0,012	465	0.01	0.55	0.70
K0,022	572	0.01	0.63	0.72
KDT2	573	0.03	0.65	0.71
KDT4	562	0.01	0.68	0.73
KDT8	636	0.02	0.72	0.87
KMt	528	0.01	0.65	0.70
KEt	576	0.01	0.71	0.79
KPr	531	0.01	0.65	0.75
KAT12	562	0.01	0.62	0.73
KAT18	462	0.01	0.56	0.69
KAT30	432	0.01	0.52	0.65
KHT12	532	0.02	0.57	0.67
KHT18	546	0.01	0.61	0.72
KHT30	624	0.01	0.72	0.81

S_{BET} = specific area; V_T = total pore volume; V_{Meso} = mesopores volume; V_{Micro} = micropores volume.

The discussions of the textural properties were approached first for the parameters of specific area (S_{BET}), microporous volume, mesoporous volume (V_{MESO}) and total pore volume (V_T), and later for the PSD, with the parameters average pore diameter and wall thickness (W_T).

The results obtained in the modification on molar proportion of organic template reveals that the reduction of P123 concentration (K0.012 sample) promoted a decrease of more than 130 $m^2.g^{-1}$ in specific area, and 0.13 $cm^3.g^{-1}$ in the values of the mesoporous volume. These variations can be explained by the decrease of micelles formed in the initial phase of the synthesis, reducing the interaction of the Si–OH groups with the organic phase. As a consequence, there was a restriction in the quantity of channels formed, resulting in a lower value of mesoporous volume, when compared with the standard KIT-6 material, since both have 3D cubic structure. The significant value of total volume for the K0.012 sample can be explained by the adsorbed volume at pressures above 0.8 P/P_0, which volume can be attributed to the interparticle spaces. For the K0.022 sample, it denotes that the increase in concentration of P123 provided the formation of a material with unidirectional hexagonal structure like as SBA-15, with relevant textural properties and similar to materials described in the literature [39,40]

Analyzing the results of the modification on the dissolution time of P123 for the materials with 3D cubic structure, it could be noted that a remarkable a proportional increase in the specific area occurred as the time increases (KDT4, KIT-6 and KDT-8 samples, respectively). This proportional difference can be explained by the dissolution time of P123, which is necessary for the organic–inorganic phase interactions to occur in a more efficient way, reflecting in the specific area. The data for the mesoporous volume and total volume of the KDT8 sample showed a significant increase when compared to the standard KIT-6, while for the KDT4 sample such values were similar (see Table 2). Such results complement the discussions on the X-ray diffractogram, where it can be determined that the increase on micellar formation positively influences the textural properties of the materials obtained. Both materials had textural properties similar to those found in the literature [5,41]. Analyzing the data for the KDT2 sample, relative to the unidirectional hexagonal structure of the SBA-15 type, relevant textural properties emphasizing the high mesoporous and microporous volume can be noted.

The modification of the type of alcohol promoted some changes in the textural properties of the materials, besides the targeting of the unidirectional hexagonal structure already described in the XRD analysis. This denotes a growing behavior of the specific area and the total and mesoporous volumes when comparing the KMt and KEt samples. However, for the sample synthesized with propanol (KPr sample), it denotes decreasing behavior, corroborating XRD observations and fortifying the idea that the size of the carbonic chain directly interferes on the synthesis mechanism. It is possible to note through these results that the disorder seen in the XRD analysis for the KPr sample can also be observed in the textural properties, which can indicate that this material shows a structure with disorderly channels. However, it should be emphasized that, even if the desired cubic materials were not obtained, the results are quite interesting. Highly organized mesoporous materials of SBA-15 type were obtained, with relevant textural properties and with a reduction of 24 h in synthesis time, when compared to the typical synthesis of these materials [28]. This denotes the quality of the material obtained, when comparing the samples in question with some found in the literature, like those reported by Kruk et al. [33] whereby the author synthesized materials of SBA-15 type with pore diameters between 5.8 and 7.5 nm and pore volumes between 0.4 and 0.8 $cm^3.g^{-1}$.

From the materials obtained through the modification of aging time in the synthesis gel (KAT12, KAT18 and KAT30 samples), it was observed that the variation in time significantly affected the textural properties. When evaluating the materials with 3D cubic structure, the values of specific area (S_{BET}), mesoporous volume (V_{MESO}) and total volume (V_T) were lower (see Table 2) for both 18 h and 30 h when compared to the 24 h for the standard KIT-6. In this case, the aging time defines an optimal equilibrium point of 24 h to promote the hydrolysis of the silica source (TEOS), and later condensation of the inorganic phase interacting with the organic phase, generating exceptional textural properties for the 3D cubic structure.

It can be observed that the variation in the time of the thermal treatment significantly influences the textural properties of the samples (KHT12, KHT18 and KHT30), and keeps the target for the 3D cubic structure. It denotes the proportional growth of values for the S_{BET}, V_{MESO} and V_T parameters when the thermal treatment time increases (see Table 2). This behavior occurs because the longer the treatment time the greater the condensation rate of the silanol groups (Si-OH), directly influencing the formation of the inorganic structure.

All the materials presented microporous volume varying between 0.01 and 0.03 $cm^3.g^{-1}$. In some specific cases, a small difference in the microporous volume is observed, such as in the case of the KHT12, KHT18 and KHT30 samples. However, the *α-plot* method does not allow a more in-depth analysis of these data, requiring more efficient methods, such as CO_2 adsorption at high pressures [42].

Table 4 shows the values of average pore diameter calculated with BJH, NLDFT and with VBS from the adsorption branch to compare the methods each other, and the values of pore diameter calculated from the desorption branch calculated by the VBS method. Figure 5 representatively shows the PSD of the standard KIT-6 and KDT2 samples.

Table 4. Average values of pore diameter through the Barrett–Joyner–Halanda (BJH), Villarroel–Bezerra–Sapag (VBS) and non-local density functional theory (NLDFT) techniques of all samples with ordered structures.

Samples	W_{BJH}/nm	W_{VBS}/nm	W_{NLDFT}/nm	W_{VBS}/nm	Wt/nm
		Adsorption		Desorption	
KIT-6	7.2	6.8	9.6	7.0	5.8
K0.012	6.8	7.0	9.1	6.6	6.9
K0.022	6.6	6.8	8.7	6.7	5.7
KDT2	6.7	6.8	9.1	6.6	6.1
KDT4	6.9	7.0	9.2	6.5	4.6
KDT8	7.2	7.2	9.6	7.2	5.4
KMt	6.5	6.7	8.7	6.5	6.2
KEt	6.8	7.0	9.1	6.7	6.5
KPr	6.5	6.7	8.9	6.7	5.0
KAT12	6.7	6.7	8.3	6.6	5.7
KAT18	6.9	7.3	8.2	7.5	5.2
KAT30	6.0	7.5	8.2	7.3	5.6
KHT12	6.1	6.5	8.2	6.6	5.9
KHT18	6.2	6.6	8.3	6.8	5.8
KHT30	7.0	6.9	9.0	7.0	5.6

W_{BJH} = pore diameter using the BJH method. W_{NLDFT} = pore diameter using the NLDFT method. W_{VBS} = pore diameter using the VBS method. Wt = wall thickness calculated from: $a_{Cub} - W_{VBS}$ (cubic structures) and $a_{Hex} - W_{VBS}$ (hexagonal structures).

The PSD is very important to define the pore volumes for each pore size and, thus, to obtain the values of the average pore diameter of nanoporous materials. From the N_2 adsorption–desorption experimental isotherm data at 77 K, it was possible to apply microscopic (based on molecular theories) and macroscopic (based on capillary condensation) methods. According to some [24,43,44], the most used microscopic method is the NLDFT, and the macroscopic methods are BJH and VBS, these last two based on the Kelvin equation.

In this work, the materials obtained with cylindrical pores show reversible isotherms with filling/emptying mechanism and hysteresis loops that are well defined. For this reason, the adsorption branch isotherm was selected to compare the methods with each other. From the evaluation of the PSD of representative samples (Figure 6) and the values of average pore diameter of all samples (Table 4), the NLDFT model overestimated the value of the pore size, while the values of the BJH and VBS methods were very close. This behavior suggests that the NLDFT model was not applied efficiently, once it takes another predetermined kernel to best fit the behavior of the samples. Among the macroscopic methods, the VBS method is the most accepted because this method was improved by adding a correction term to the original Kelvin equation, providing more accurate values for the pore size. Taking this into account, all discussion about the PSD and average pore diameter were carried out based on the VBS method applied in the desorption branch. This is because, according to the authors [24,44], the desorption branch reflects the transition of the equilibrium phase, which supports the macroscopic methods that use the Kelvin equation.

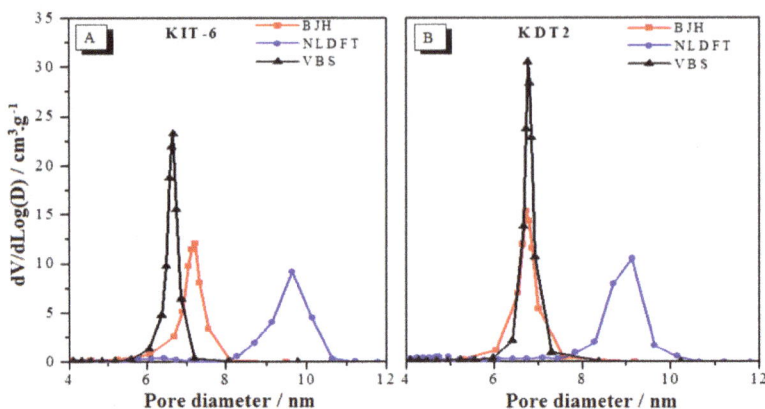

Figure 6. Pore size distribution (PSD) of the (**A**) standard KIT-6 and (**B**) KDT2 samples.

From the data of the K0.012 and K0.022 samples obtained, there can be seen a similarity in the values of pore diameter when compared to the standard KIT-6. Such a fact demonstrates that the variation of the amount of organic template did not influence in an expressive way the pore size of the materials. However, when analyzing the values of wall thickness, a difference of 1.1 nm between the K0.012 sample and the standard KIT-6 can be noted. This difference is directly related to the values of cubic parameters observed (see Table 2). In consequence, such behavior is reflected in the wall thickness. Thus, the wall thickness values contribute to the explanation of the synthesis mechanism and the specific area values for such materials, as cited previously.

Evaluating the dissolution time for the materials with 3D cubic structures, the KDT4 and KDT8 samples showed proportional differences for pore diameter, when compared to the standard KIT-6. The values of pore diameter exhibited a difference of 0.4 nm (values of 6.5, 6.9 and 7.3 nm) between the KDT4, KIT-6 and KDT8 samples, respectively. This behavior indicates an increase in the average pore size in function of the dissolution time of the organic template. Analyzing the values of wall thickness, it denotes a significant decrease of 1.2 nm on the KDT4 sample, when compared to the standard KIT-6. This data correlates the values observed for the cubic parameter (a_{cub}) and, in consequence, specific area (S_{BET}), corroborating with the synthesis mechanism cited. The KDT8 sample was compared with the standard KIT-6 and it is worth mentioning that, despite the values of wall thickness being similar, the KDT8 sample demonstrated superior textural properties (specific area, mesoporous volume, total volume and pore diameter) with the addition of 2 h in the synthesis process.

Evaluating the pore diameter data for the KMt, KEt and KPr materials with hexagonal structure, this shows very similar behavior for each one. On the other hand, a very irregular behavior was seen when comparing the data obtained for wall thickness of these materials. A similarity between the KMt and KEt samples (6.2 and 6.5 nm, respectively) can be noted, while the KPr sample shows a significantly inferior value (about 1.3 nm). This behavior relates to the values of interplanar distance and mesoporous parameter for the unidirectional hexagonal structure. This set of facts reveals that the type of alcohol used in the synthesis acts only as a co-solvent for the formation of the SBA-15 structure, slightly modifying the textural properties, but not significantly interfering in the micellar formation.

The samples that displayed 3D cubic structure in the modification of aging time of the synthesis gel (KAT18 and KAT30 samples) presented values of average pore diameter and wall thickness similar to the standard KIT-6 material. Such a result suggests that this synthesis parameter does not significantly interfere in the formation and micellar organization. However, as discussed earlier, there is an optimal equilibrium time of 24 h to promote better textural properties values (S_{BET}, V_{MESO} and V_T), based in the stages of hydrolysis of the silica and condensation of the inorganic phase. From this, it is of great relevance to point out that the aging time of synthesis gel favors the 3D cubic structure for the KIT-6 material starting from 18 h of procedure, reaching the best textural properties in 24 h.

Analyzing the samples obtained in the variation of thermal treatment time, this denotes a small increase in the average pore diameter (0.4 nm) and a slight decrease in wall thickness (0.3 nm), when comparing the shortest time studied (KHT12 sample) with the longest (KHT30 sample). These results reveal that the modification of thermal treatment time, in which occurs the condensation of the silanol groups, does not significantly influence the micellar formation and consequently the average pore diameter. However, the thermal treatment time acts considerably in the synthesis mechanism, increasing the values of the textural properties (specific area, mesoporous volume and total volume), when compared to the 12 and 30 h times (see Table 2). Nevertheless, it is important to emphasize that, starting with 12 h of thermal treatment, the 3D cubic structure is formed with relevant textural properties. With an increase in 6 h at a time between the KIT-6 and KHT30 samples, the 3D cubic structure reflects similar textural properties.

Figure 7 exhibits TEM micrographs of KDT4 and KHT12 representative samples.

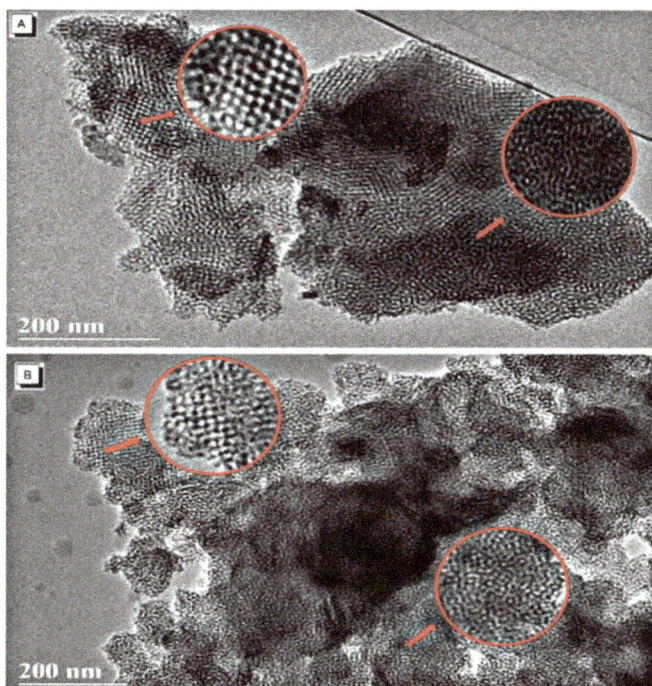

Figure 7. Representative transmission electron microscope (TEM) images of (**A**) KDT4 and (**B**) KHT12 samples.

Appl. Sci. **2018**, *8*, 725

The images of the KDT4 and KHT12 samples are presented in a representative way for the materials that showed a 3D cubic structure. Analyzing the TEM micrographies of these materials, it was observed along the axis [001] that the two samples showed well-defined pore organization. When analyzing the image along the axis [010], it can be clearly noticed that the channel curvatures refer to the type of bi-continuous channels respective to KIT-6 mesoporous materials. Such behavior can also be visualized in studies described in the literature [9,45]. These results corroborate those found in the XRD technique and N_2 adsorption–desorption isotherms.

Through this technique, the pore diameter of these materials was also measured. The KDT4 and KHT12 samples presented average values of 6.4 ± 1.23 and 5.9 ± 0.95 nm for the pore diameters, respectively. These values are very similar to those found in the pore size distribution (PSD), calculated by the VBS method in the desorption branch through N_2 adsorption–desorption isotherms. From this, it is important to emphasize two points: the employment of the VBS method for measuring the pore diameter by the pore size distribution (PSD) was precisely appropriated; and the modification in some parameters, reducing the time at synthesis procedure, promoted KIT-6 materials with a well-defined 3D cubic structure.

4. Conclusions

The KIT-6 mesoporous material have been efficiently synthesized through a typical procedure, reported in the literature. Based on this procedure, modifications to the synthesis parameters have been presented and show a significant influence on the structure and textural properties in the ordered mesoporous materials obtained. For the formation of *Ia3d* cubic structures, characteristic of the KIT-6 material, concentrations of P123 below 0.017 M are required. The alcohol probably acts on the generation of the micelle, and the size of its carbonic chain possibly interferes in the formation of the porous system and ordered structure of the synthesized mesoporous materials. In order to obtain the 3D cubic structure, a longer synthesis time is required, otherwise SBA-15 or any ordered structures are formed. In other words, below 4 h and 18 h for P123 dissolution and aging gel, respectively, the SBA-15 ordered structure is generated. The heat treatment time has a significant influence on the textural properties of the synthesized materials. Thus, this parameter revealed that the longer the time employed, the better the materials that are produced, by demonstrating elevated specific area, high volume and pore diameter. The data obtained have confirmed the influence of the synthesis parameters in the formation, organization and textural properties of mesoporous material like KIT-6. However, it is important to emphasize that the cubic structure of KIT-6 can be obtained by reducing the concentration and dissolution time of P123, and times of heat treatment and aging, when compared to the synthesis reported in the literature. The modifications carried out in the synthesis procedure have resulted in materials with different characteristics, which makes it possible to employ them in applications in the areas of catalysis and adsorption.

Supplementary Materials: The following are available online at http://www.mdpi.com/2076-3417/8/5/725/s1, Figure S1: Curves of (A) TG and (B) DTG of samples with cubic structures, Figure S2: Curves of (A) TG and (B) DTG of samples with hexagonal structures, Figure S3: FTIR spectra of all samples with ordered structures.

Author Contributions: F.R.D.F. and E.L.F.L. performed the experiments. F.R.D.F., F.G.H.S.P., A.G.D.S. and V.P.S.C. wrote the paper and analyzed the data. L.D.S., V.P.S.C. and A.G.D.S. supervised the whole work.

Acknowledgments: The authors thanks to the CAPES/Brazil for the financial support, and the Magnetic and Optical Analysis Laboratory (LAMOp/State University of Rio Grande do Norte), and GERATEC-PPGQ-CCN/State University of Piauí, for the analysis performed.

Conflicts of Interest: The authors declare no conflict of interest. The founding sponsors had no role in the design of the study; in the collection, analyses, or interpretation of data; in the writing of the manuscript, and in the decision to publish the results.

References

1. Prabhu, A.; Kumaresan, L.; Palanichamy, M.; Murugesan, V. Synthesis and characterization of aluminium incorporated mesoporous KIT-6: Efficient catalyst for acylation of phenol. *Appl. Catal. A Gen.* **2009**, *360*, 59–65. [CrossRef]

2. Qian, L.; Ren, Y.; Liu, T.; Pan, D.; Wang, H.; Chen, G. Influence of KIT-6's pore structure on its surface properties evaluated by inverse gas chromatography. *Chem. Eng. J.* **2009**, *213*, 186–194. [CrossRef]

3. Kalsabi, R.J.; Mosaddegh, N. Pd-poly(*N*-vinyl-2-pyrrolidone)/KIT-6 nanocomposite: Preparation, structural study, and catalytic activity. *C. R. Chim.* **2012**, *15*, 988–995. [CrossRef]

4. Dou, B.; Hu, Q.; Li, J.; Qiao, S.; Hao, Z. Adsorption performance of VOCs in ordered mesoporous silicas with different pore structures and surface chemistry. *J. Hazard. Mater.* **2011**, *186*, 1615–1624. [CrossRef] [PubMed]

5. Boulaoued, A.; Fechete, I.; Donnio, B.; Bernard, M.; Turek, P.; Garin, F. Mo/KIT-6, Fe/KIT-6 and Mo-Fe/KIT-6 as new types of heterogeneous catalysts for the conversion of MCP. *Microporous Mesoporous Mater.* **2012**, *155*, 131–142. [CrossRef]

6. Kumaseran, L.; Prabhu, A.; Palanichamy, M.; Murugesan, V. Mesoporous Ti-KIT-6 molecular sieves: Their catalytic activity in the epoxidation of cyclohexene. *J. Taiwan Inst. Chem. Eng.* **2010**, *41*, 670–675. [CrossRef]

7. Falahati, M.; Ma'mani, L.; Saboury, A.A.; Shafiee, A.; Foroumadi, A.; Badiei, A.B. Aminopropyl-functionalized cubic Ia3d mesoporous silica nanoparticle as an efficient support for immobilization of superoxide dismutase. *Biochim. Biophys. Acta* **2012**, *1814*, 1195–1202. [CrossRef] [PubMed]

8. Karthikeyan, G.; Pandurangan, A. Post synthesis alumination of KIT-6 materials with Ia3d symmetry and their catalytic efficiency towards multicomponent synthesis of 1*H*-pyrazolo [1,2-]phthalazine-5,10-dione carbonitriles and carboxylates. *J. Mol. Catal. A Chem.* **2012**, *361*, 58–67. [CrossRef]

9. Kleitz, F.; Choi, S.H.; Ryoo, R. Cubic Ia3d mesoporous sílica: Synthesis and replication to platinum nanowires, carbon nanorods and carbon nanotubes. *Chem. Commun.* **2003**, *17*, 2136–2137. [CrossRef]

10. Zhao, Y.; Chen, X.; Yang, C.; Zhang, G. Mesoscopic Simulation on Phase Behavior of Pluronic P123 Aqueous Solution. *J. Phys. Chem. B* **2007**, *111*, 13937–13942. [CrossRef] [PubMed]

11. Wang, Y.; Zhang, F.; Wang, Y.; Ren, J.; Li, C.; Liu, X.; Guo, Y.; Guo, Y.; Lu, G. Synthesis of length controllable mesoporous SBA-15 rods. *Mater. Chem. Phys.* **2009**, *115*, 649–655. [CrossRef]

12. Impéror-Clerc, M.; Manet, S.; Grillo, I.; Durand, D.; Khodakov, A.; Zholobenko, V. Study of the mechanisms and kinetics of the synthesis of mesoporous materials from micelles of tri-block copolymers. *Stud. Surf. Sci. Catal.* **2008**, *174*, 805–810. [CrossRef]

13. Guan, L.S.; Nur, H.; Endud, S. Bimodal pore size mesoporous mcm-48 prepared by post-synthesis alumination. *J. Phys. Sci.* **2006**, *17*, 65–75.

14. Chareonpanich, M.; Nanta-Ngern, A.; Limtrakul, J. Short-period synthesis of ordered mesoporous silica SBA-15 using ultrasonic technique. *Mater. Lett.* **2007**, *61*, 5153–5156. [CrossRef]

15. Cao, L.; Dong, H.; Huang, L.; Matyjaszewski, K.; Kruk, M. Synthesis of large pore SBA-15 silica using poly(ethylene oxide)-poly(methyl acrylate) diblock copolymers. *Adsorption* **2009**, *15*, 156–166. [CrossRef]

16. Galarneau, A.; Cambon, H.; Renzo, F.D.; Ryoo, R.; Choi, M.; Fajula, F. Microporosity and connections between pores in SBA-15 mesostructured silicas as a function of the temperature of synthesis. *New J. Chem.* **2002**, *7*, 73–79. [CrossRef]

17. Kang, K.; Rhee, H. Synthesis and characterization of novel mesoporous silica with large wormhole-like pores: Use of TBOS as silicon source. *Microporous Mesoporous Mater.* **2005**, *84*, 34–40. [CrossRef]

18. Iglesias, J.; Melero, J.A.; Sainz-pardo, J. Direct synthesis of organically modified Ti-SBA-15 materials. *J. Mol. Catal. A Chem.* **2008**, *291*, 75–84. [CrossRef]

19. Qiang, Z.; Gurkan, B.; Ma, J.; Liu, X.; Guo, Y.; Cakmak, M.; Cavicchi, K.A.; Vogt, B.D. Roll-to-roll fabrication of high surface area mesoporous carbon with process-tunable pore texture for optimization of adsorption capacity of bulky organic dyes. *Microporous Mesoporous Mater.* **2016**, *227*, 57–64. [CrossRef]

20. Deng, G.; Zhang, Y.; Ye, C.; Qiang, Z.; Stein, G.E.; Cavicchi, K.A.; Vogt, B.D. Bicontinuous mesoporous carbon thin films *via* an order-order transition. *Chem. Commun.* **2014**, *50*, 12684–12687. [CrossRef] [PubMed]

21. Brunauer, B.S.; Emment, P.H.; Teller, E. Adsorption of gases in multimolecular layers. *J. Am. Chem. Soc.* **1938**, *60*, 309–319. [CrossRef]

22. Rouquerol, J.; Llewellyn, P.; Rouquerol, F. Is BET equation applicable to microporous adsorbents? *Stud. Surf. Sci. Catal.* **2007**, *160*, 49–56.

23. Rouquerol, F.; Rouquerol, J.; Sing, K.S.W.; Llewellyn, P.; Maurin, G. *Adsorption by Powders and Porous Solids*, 2nd ed.; Methodology and Applications, Academic Press: San Diego, CA, USA, 2014; ISBN 978-0-08-097035-6.

24. Villarroel-Rocha, J.; Barrera, D.; Sapag, K. Introducing a self-consistent test and the corresponding modification in the Barrett, Joyner and Halenda method for pore-size determination. *Microporous Mesoporous Mater.* **2014**, *200*, 68–78. [CrossRef]

25. Zukal, A.; Thommes, M.; Čejka, J. Synthesis of highly ordered MCM-41 silica with spherical particles. *Microporous Mesoporous Mater.* **2007**, *104*, 52–58. [CrossRef]

26. He, C.; Li, J.; Zhang, X.; Yin, L.; Chen, J.; Gao, S. Highly active Pd-based catalysts with hierarchical pore structure for toluene oxidation: Catalyst property and reaction determining factor. *Chem. Eng. J.* **2012**, *180*, 46–56. [CrossRef]

27. Zhao, H.; Liu, S.; Wang, R.; Zhang, T. Humidity-sensing properties of LiCl-loaded 3D cubic mesoporous silica KIT-6 composites. *Mater. Lett.* **2015**, *147*, 54–57. [CrossRef]

28. Zhao, D.; Feng, J.; Huo, Q.; Melosh, N.; Fredrickson, G.H.; Chmelka, B.F.; Stucky, G.D. Triblock Copolymer Syntheses of Mesoporous Silica with Periodic 50 to 300 Angstrom Pores. *Science* **1998**, *279*, 548–552. [CrossRef] [PubMed]

29. Wanka, G.; Hoffmann, H.; Ulbricht, W. Phase Diagrams and Aggregation Behavior of Poly(oxyethylene)-Poly(oxypropylene)-Poly(oxyethylene) Triblock Copolymers in Aqueous Solutions. *Macromolecules* **1994**, *27*, 4145–4159. [CrossRef]

30. Kim, T.W.; Kleitz, F.; Paul, B.; Ryoo, R. MCM-48-like Large Mesoporous Silicas with Tailored Pore Structure: Facile Synthesis Domain in a Ternary Triblock Copolymer-Butanol-Water System. *J. Am. Chem. Soc.* **2005**, *127*, 7601–7610. [CrossRef] [PubMed]

31. Kim, J.M.; Ryoo, R. Synthesis of MCM-48 single crystals. *Chem. Commun.* **1998**, *2*, 259–260. [CrossRef]

32. Liu, C.; Wang, X.; Lee, S.; Pfefferle, L.D.; Haller, G.L. Surfactant chain length effect on the hexagonal-to-cubic phase transition in mesoporous silica synthesis. *Microporous Mesoporous Mater.* **2012**, *147*, 242–251. [CrossRef]

33. Kruk, M.; Jeroniec, M.; Ko, C.H.; Ryoo, T. Characterization of the Porous Structure of SBA-15. *Chem. Mater.* **2000**, *12*, 1961–1968. [CrossRef]

34. Fernandes, F.R.D.; Santos, A.G.D.; Souza, L.D.; Santos, A.P.B. Síntese e caracterização do material mesoporoso sba-15 obtido com diferentes condições de síntese. *Rev. Virtual Química* **2016**, *8*, 1855–1864. [CrossRef]

35. Shukla, P.; Sun, H.; Wang, S.; Ang, H.M.; Tadé, M.O. Co-SBA-15 for heterogeneous oxidation of phenol with sulfate radical for wastewater treatment. *Catal. Today* **2011**, *175*, 380–385. [CrossRef]

36. Wang, X.Q.; Ge, H.L.; Jin, H.X.; Cui, Y.J. Influence of Fe on the thermal stability and catalysis of SBA-15 mesoporous molecular sieves. *Microporous Mesoporous Mater.* **2005**, *86*, 335–340. [CrossRef]

37. Guo, Y.H.; Xia, C.; Liu, B.S. Catalytic properties and stability of cubic mesoporous LaxNiyOz/KIT-6 catalysts for CO2 reforming of CH4. *Chem. Eng. J.* **2014**, *237*, 421–429. [CrossRef]

38. Thommes, M.; Koneko, K.; Neimark, A.V.; Oliver, J.P.; Rodirguez-reinoso, F.; Rouquerol, J.; Sing, K.S.W. Physisorption of gases, with special reference to the evaluation of surface area and pore size distribution (IUPAC Technical Report). *Pure Appl. Chem.* **2015**, *87*, 1051–1069. [CrossRef]

39. Wang, S.; Wang, K.; Dai, C.; Shi, H.; Li, J. Adsorption of Pb^{2+} on amino-functionalized core-shell magnetic mesoporous SBA-15 silica composite. *Chem. Eng. J.* **2015**, *262*, 897–903. [CrossRef]

40. Vizcaíno, A.J.; Carrero, A.; Calles, J.A. Comparison of ethanol steam reforming using Co and Ni catalysts supported on SBA-15 modified by Ca and Mg. *Fuel Process. Technol.* **2016**, *146*, 99–109. [CrossRef]

41. Hussain, M.; Akhter, P.; Russo, N.; Saracco, G. Novel Ti-KIT-6 material for the photocatalytic reduction of carbon dioxide to methane. *Catal. Commun.* **2013**, *36*, 58–62. [CrossRef]

42. Thommes, M.; Chychosz, K.A.; Neimark, A.V. Advanced physical adsorption characterization of nanoporous carbons. *Novel Carb. Adsorbents* **2012**, 107–139. [CrossRef]

43. Neimark, A.V.; Ravikovitch, P.I. Density Functional Theory of Adsorption Hysteresis and Nanopore Characterization. *Stud. Surf. Sci. Catal.* **2000**, *128*, 51–60. [CrossRef]

44. Rocha, J.V.; Barrera, D.; Sapag, K. Improvement in the Pore Size Distribution for Ordered Mesoporous Materials with Cylindrical and Spherical Pores Using the Kelvin Equation. *Top. Catal.* **2011**, *54*, 121–134. [CrossRef]
45. Wang, J.; Li, Y.; Zhang, Z.; Hao, Z. Mesoporous KIT-6 silica-polydimethylsiloxane (PDMS) mixed matrix membranes for gas separation. *J. Mater. Chem. A* **2015**, *3*, 8650–8658. [CrossRef]

© 2018 by the authors. Licensee MDPI, Basel, Switzerland. This article is an open access article distributed under the terms and conditions of the Creative Commons Attribution (CC BY) license (http://creativecommons.org/licenses/by/4.0/).

applied
sciences

MDPI

Article

Nickel Complexes Immobilized in Modified Ionic Liquids Anchored in Structured Materials for Ethylene Oligomerization

Camila A. Busatta [1], Marcelo L. Mignoni [2], Roberto F. de Souza [3,†] and Katia Bernardo-Gusmão [3,*]

[1] Departmento de Ciências Exatas e da Terra, Universidade Regional Integrada do Alto Uruguai e das Missões-URI, Av. Assis Brasil, 790, Itapagé, Frederico Westphalen 98400-000, Brazil; camilapaguilar@yahoo.com.br
[2] Departmento de Química, Universidade Regional Integrada do Alto Uruguai e das Missões-URI, Av. Sete de Setembro, 1621, Erechim 99700-000, Brazil; mignoni@uricer.edu.br
[3] LRC, Instituto de Química, Universidade Federal do Rio Grande do Sul-UFRGS, Av. Bento Gonçalves, 9500, P.O. BOX 15003, Porto Alegre 91501-970, Brazil; rfds@iq.ufrgs.br
* Correspondence: katiabg@iq.ufrgs.br; Tel.: +55-51-3308-6481
† In memoriam.

Received: 21 March 2018; Accepted: 24 April 2018; Published: 4 May 2018

Featured Application: Immobilized complexes in porous materials (β-zeolite, MCM-41 and Al-MCM-41) modified with silylated ionic liquid were prepared and the catalytic systems were tested for ethylene oligomerization under homogeneous and heterogeneous conditions using toluene and EASC.

Abstract: This work describes the study of ethylene oligomerization reactions catalyzed by nickel-β-diimine complexes immobilized on β-zeolite, [Si]-MCM-41 (Mobil Composition of Matter 41) and [Si,Al]-MCM-41 modified with an ionic liquid. XRD and N_2 adsorption and desorption analyses were used to characterize the modified supports—namely, IL-Zeoβ, IL-MCM-41 and IL-Al-MCM-41—and the data showed that material organization remained intact even after incorporation of ionic liquid. N_2 adsorption and desorption analyses suggested that ionic liquid can be confined in pores of support materials. Catalytic properties of synthesized materials were tested under different conditions. The following parameters were varied: Al/Ni molar ratio, temperature, pressure and catalyst loading. The homogeneous catalysts were more active but less selective in ethylene oligomerization, relative to heterogeneous ones, which can be attributed to the effect of confinement suffered by catalyst within channels of the support materials. NiIL-Zeoβ complexes were active, with activities greater than 23 s^{-1} and selectivities higher than 80% for butenes, including more than 85% of 1-butene. On the other hand, the NiIL-MCM-41 system was less active than NiIL-Zeoβ complexes, with activities above 1 s^{-1} with 100% selectivity for butenes (96% in 1-butene). NiIL-Al-MCM-41 system was more active than NiIL-MCM-41 system and showed an activity of 2.3 s^{-1} with 90% selectivity in 1-butenes.

Keywords: ionic liquid; β-diimine; nickel; heterogenized; β-zeolite; MCM-41; oligomerization

1. Introduction

The oligomerization of ethylene is an important industrial reaction, and megatons of α-olefins are produced in this way every year. Depending on the chain length of the alkene, these materials are used for manufacturing various products, ranging from plastics to lubricants and linear low-density polyethylene (LLDPE). Among the industrial processes used for ethylene oligomerization using homogeneous catalysis, many processes were developed by companies such as Shell (Shell High

Olefins Process (SHOP)) [1] and the IFP—Institut Français du Pétrole (Dimersol and Alfabutol) [2,3]. These systems allow the production of products with high activities and selectivities; however, the separation of the products from the reaction medium makes catalyst recycling a challenge. The use of heterogeneous catalytic systems may be an interesting way to address the issue of separating the catalyst from the reaction products, increase the catalysts' resistance and facilitate its reuse [4].

Currently, a wide variety of molecular sieves are available, including zeolites, and these materials have become important as catalysts and adsorbents [5–7]. Molecular sieves are most commonly used in processes that employ relatively small molecules due to the sizes of their pores, which are approximately 8 Å [8]. The β-zeolite described by Mobil Oil Corporation is an example of a zeolite structure that is applicable in such a situation. The zeolite is a β-zeolite, which has a three-dimensional system of channels and micropores (diameters of 7.5 Å) that are circumscribed by rings of 12 tetrahedra that can be directly synthesized with a considerably high Si/Al ratio. Its high acidity, thermal stability, and hydrothermal stability and the ease by which relatively large molecules can diffuse through the channels make this a very interesting material from the perspective of developing new catalytic zeolites.

Other supports used with the same propose are the mesoporous MCM-41 (Mobil Composition of Matter 41) solids. These supports represent a new platform because their morphological characteristics and pore diameters of between 2 and 50 nm allow the introduction of compounds with greater steric hindrance into the cavities. The solids [Si]-MCM-41 and [Si,Al]-MCM-41 [9] have high thermal stability and a hexagonal arrangement of pores with linear and parallel channels and are constructed with a beehive-type silica matrix. Use of MCM-41 as support of several metal complexes that are catalysts of ethylene, propylene or 1-hexene oligomerization reactions was described in the literature [6,10–14]. In all cases, the reported oligomer selectivity was affected by the presence of MCM-41.

We have used ionic liquids as solvents in several types of catalytic reactions (biphasic catalysis) [15,16]. Recently, a new type of tether was described in the literature in which the complex is anchored to the ionic liquid that is supported on mesoporous silica; this technique has been employed in hydroformylation [17–19], carbonylation [20], hydrogenation [21], Heck [22] and epoxidation [23] reactions.

In this work, a nickel-β-diimine complex was immobilized on β-zeolite, [Si]-MCM-41 and [Si,Al]-MCM-41 modified with the tetrafluoroborate of 1-(3-trimetoxisililpropil)-3-methylimidazolium ionic liquid in the presence of the tetrafluoroborate of 1-butyl-3-methylimidazolium (BMI.BF4). The catalyst obtained was used in ethylene oligomerization reactions.

2. Materials and Methods

All syntheses were performed under an argon atmosphere using standard Schlenk techniques. Ethylene (White Martins 99.99%) was used for the catalytic tests. The cocatalyst ethylaluminum sesquichloride (EASC) was used after being diluted in toluene (10% v/v). Characterization of the compounds was done using gas chromatography (GC), nuclear magnetic resonance (NMR), elemental analysis, N_2 adsorption, infrared (IR) spectroscopy, X-ray diffraction (XRD), and atomic absorption spectroscopy (AAS), as detailed in the following. GC was performed in a Varian Star 3400 CX chromatograph with a flame ionization detector, equipped with a Petrocol DH capillary column (methyl silicone, 100 m in length, 0.25 mm ID, 0.5 μm film thickness). The temperature was initially maintained at 36 °C for 15 min, followed by heating to 250 °C at a rate of 5 °C/min. ^1H- and ^{13}C-NMR spectroscopy employed a Varian 300 NMR spectrometer, operating at 300 MHz for ^1H-NMR and 75 MHz for ^{13}C-NMR, using CDCl$_3$ as solvent. Elemental analysis used a Perkin Elmer M CHN/O Model 2400 system. N_2 adsorption utilized a Quantachrome Nova 2200e model system in conjunction with the BET method to determine the specific surface area of samples previously treated at 80 °C by 3 h. Vibrational spectroscopy used a Shimadzu infrared spectrometer. XRD was performed using a diffractometer from Siemens (model D500) with CuKα radiation (λ = 1.54 Å). For AAS, done in a spectrometer from Perkin Elmer (A. Analyst 200), 20 mg of samples were prepared with 2 mL of HCl,

6 mL of HNO_3, 5 mL of HF, adding the mixture to Teflon autoclaves, subsequently using a digester for 10 h at 150 °C. After cooling, the samples were diluted to 50 mL.

2.1. Ligand 2-(phenyl)amino-4-(phenyl)imino-2-pentene

An argon-filled glass flask was charged with 30 mL of toluene, 10 mL of acetylacetone (100 mmol) and 18 mL of aniline (200 mmol). This mixture was cooled in an ice bath and then 8.3 mL of concentrated hydrochloric acid was slowly added to the flask. After 24 h, a precipitate was obtained that was then isolated by filtration and washed with hexane. The solid was neutralized and extracted by adding 8 mL of dichloromethane, 50 mL of distilled water and 20 mL of saturated solution of sodium carbonate. Next, a pear-shaped separation flask was used to extract the organic phase. The solution was concentrated under reduced pressure, and the ligand was crystalized from methanol and left in the freezer. We obtained 9.13 g of the ligand (36.5 mmol, 36.5% yield) as a yellow solid. The ligand was characterized by ^{1}H-NMR spectroscopy ($CDCl_3$, 300 MHz, Room temperature, δ in ppm): 12.7 (s, 1H, H_9), 7.3–6.9 (m, 10H, H_4, H_5, H_6, H_7, H_8), 4.9 (s, 1H, H_2), 2 (6H, H_1, H_3). Anal. Calc for $C_{17}H_{18}N_2$: C, 81.56; H, 7.25; N, 11.19. Found: C, 81.24; H, 7.41; N, 11.35.

2.2. Synthesis of the Bis-(acetonitrile)dibromidonickel(II) Adduct [24]

To a 250 mL Schlenk tube, 120 mL of acetonitrile and 2.26 g of $NiBr_2$ (anhydrous) were added. This suspension remained under reflux at 80 °C until a dark blue color was obtained. The final solution was concentrated to approximately 40 mL under reduced pressure, and the resulting solid was filtered through a fritted filter Schlenk tube, washed with 30 mL of acetonitrile and then dried under flowing argon.

2.3. Synthesis of the 1,5-Bisphenyl-2,4-pentanediimine-dibromidonickel(II) Complex

The complex was synthesized as described in the literature [25]. To a 250 mL Schlenk tube, the diimine ligand (0.77 g, 3.08 mmol) (which was synthesized as described in Section 3.1) and 40 mL of dichloromethane were added. To this solution, the *bis*-(acetonitrile)dibromidonickel(II) adduct (0.914 g, 3.04 mmol) was added. The reaction mixture was magnetically stirred for 4 days at room temperature. After the reaction was complete, the solution was concentrated to approximately 15 mL and filtered through a fritted filter Schlenk tube before the solvent was evaporated under vacuum to obtain the resulting solid nickel complex (light purple). We isolated 0.48 g of the complex (33% yield). Anal. Calc for $C_{19}H_{24}Br_2N_2Ni$: C, 45.74; H, 4.85; N, 5.61 Found: C, 44.76; H, 4.12; N, 6.01.

2.4. Synthesis of 1-(3-Trimethoxysilylpropyl)-3-methylimidazolium Chloride

The procedure used for the preparation of the ionic liquid was similar to the methods described in the literature [26]. 1-Methylimidazole (3.36 g, 40 mmol) and 7.92 g of 3-chloropropyltrimethoxysilane (40 mmol) were added to a Schlenk tube under argon. The reaction mixture was stirred for 24 h at 95 °C. A viscous oil formed after heating. This oil was used immediately in the next step. The assignments of the ^{1}H-NMR and ^{13}C-NMR signals of 1-(3-trimetoxisililpropil)-3-methylimidazoliumchloride are as follows: ^{1}H-NMR ($CDCl_3$): δ (ppm) = 0.65 (t, 2H), 2.0 (m, 2H), 3.6 (s, 9H), 4.1 (s, 3H), 4.36 (t, 2H), 7.54 (s, 1H), 7.83 (s, 1H), 10.5 (s, 1H); ^{13}C-NMR ($CDCl_3$): δ (ppm) = 5.5 ($SiCH_2$), 24.0 (CH_2), 36.4 (CH_3N), 51.2 (OCH_3), 76.9 (CH_2N), 121.7, 123.4, 137.2.

2.5. Synthesis of 1-(3-Trimethoxysilylpropyl)-3-methylimidazoliumhexafluoro-phosphate

To a solution of the above compound (3.2 g, 10 mmol) in 15 mL of acetone, sodium tetrafluoroborate (1.15 g, 10.5 mmol) was added. This mixture was stirred at room temperature for 3 days. The solid was then removed by filtration, and the solvent was evaporated. The assignments of the ^{1}H-NMR and ^{13}C-NMR signals of the tetrafluoroborate 1-(3-trimetoxisililpropil)-3-methylimidazolium are as follows: ^{1}H-NMR ($CDCl_3$), δ (ppm) = 0.65 (t, 2H), 2.0 (m, 2H), 3.6 (s, 9H), 3.9 (s, 3H), 4.2 (t, 2H), 7.34 (s, 1H),

7.39 (s, 1H), 8.90 (s, 1H); ^{13}C-NMR (CDCl$_3$), δ (ppm) = 6.1 (SiCH$_2$), 24.3 (CH$_2$), 36.7 (CH$_3$N), 51.7 (CH$_3$O), 52.1 (CH$_2$N), 122.0, 124.2, 137.1.

2.6. Synthesis of the Supports

β-Zeolite, with a ratio Si/Al = 20, was synthesized as describe in the literature [27]. Before use, the β-zeolite was calcined at 600 °C for 6 h at a heating rate of 3 °C/min to remove the ionic liquid, which had been used as a structure-directing agent and was present in the cavities.

The mesoporous supports, [Si]-MCM-41 and [Si,Al]-MCM-41, were synthesized according to the route proposed by Corma [28]. Finally, the obtained samples were calcined in air under static conditions at 540 °C for 6 h at a heating rate of 0.5 °C/min, with 60 min holds at 150 and 350 °C.

2.7. Synthesis of Silylated Ionic Liquid-Modified Supports

Ionic liquid (6.5 mmol) (synthesized as described in Section 2.4) was dissolved in chloroform (50 mL) and mixed with 1 g of support (previously dried under vacuum and heated to 180 °C overnight). The mixture was heated to 65 °C under reflux and stirred for 26 h. The mixture was cooled to room temperature and filtered, and the solid was washed with chloroform (50 mL) and diethyl ether (50 mL). Afterwards, the solid was dried under reduced pressure.

The synthesized materials were named IL-Zeoβ, IL-MCM-41 and IL-Al-MCM-41. The incorporation of the silylated ionic liquid onto the materials is outlined in Figure 1.

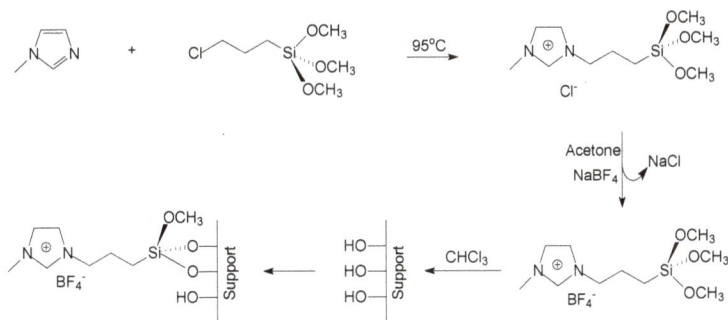

Figure 1. Scheme to synthesis of ionic liquid-modified supports.

2.8. Immobilization of the Nickel-β-Diimine Complex onto the Modified Supports

To a Schlenk tube, the nickel-β-diimine complex (0.30 mmol) dissolved in 10 mL of dichloromethane, 300 mg of ionic liquid BMI.BF$_4$ and 1 g of modified support (synthesized in the previous step) was added. The mixture was stirred for 1 h at room temperature and then the solvent was evaporated. The synthesized materials are named NiIL-Zeoβ, NiIL-MCM-41 and NiIL-Al-MCM-41.

2.9. Catalytic Tests

Ethylene oligomerization reactions were performed in triplicate, under both homogeneous and heterogeneous conditions, and the results were compared. The catalytic reactions were performed in a 200 mL double-walled glass reactor, equipped with a magnetic stirrer, a thermocouple, and a continuous feed of ethylene. The reaction temperature was controlled by a thermostatic circulation bath. All reactions employed 30 mL of toluene as the solvent and a solution of EASC as the cocatalyst. The amounts of catalyst and cocatalyst, as well as temperature and pressure, were optimized. After each reaction, the reactor was cooled with a mixture of ethanol and liquid N$_2$ (to reach temperatures up to −40 °C). After cooling, the reaction mixture was poured into a vial for later analysis, and eventual quantification by gas chromatography using isooctane as an internal standard. The catalytic tests evaluated activity (expressed by turnover frequency (TOF), s^{-1}) and selectivity (% of butenes and % of 1-butene in fraction C$_4$) as shown in Table 1.

Table 1. Expressions used for calculations of activity and selectivity of the oligomers.

Turnover Frequency (TOF)	$\dfrac{\text{mol of ethylene converted}}{\text{mol of nickel} \times \text{time (s)}}$
Selectivity (% butenes)	$\dfrac{\text{butenes mass}}{\sum \text{oligomers mass}}$
Selectivity (% 1-butene)	$\dfrac{\text{1-butene mass}}{\sum \text{butenes mass}}$
Selectivity (% internal butenes)	$\dfrac{\text{Internal butenes mass}}{\sum \text{butenes mass}}$

Catalyst-recycling tests were performed with NiIL-Zeoβ. After the first reaction, the products were removed from the reactor through a cannula and collected in a cooled Schlenk tube. Through this technique, we could isolate the NiIL-Zeoβ catalyst from the reactor while maintaining the argon atmosphere. Recycling was accomplished by adding another 30 mL of toluene and re-adding the same volume of EASC cocatalyst as in the first cycle.

3. Results and Discussion

First, we present the results of the characterization of the materials; next, we discuss the results of the catalytic oligomerization tests in detail.

3.1. Materials Characterization

In the infrared spectra of the supported complexes (Figure 2), there are specific bands at approximately 1098, 794 and 448 cm^{-1} that can be attributed to the vibrations of the mesoporous structure (Si–O), a band at 547 cm^{-1} that can be assigned to the characteristic absorption of β-zeolite and mesoporous materials, and a broad band at approximately 3409 cm^{-1} that can be assigned to silanol groups (SiO–H). The bands at 3154 and 3098 cm^{-1} are assigned to the CH vibrations of the imidazole aromatic ring and the CH stretching of the alkyl groups belonging to the silylating agent. The bands at 2960 and 2870 cm^{-1} in the spectra of the supported complexes are assigned to symmetric and asymmetric stretches of the CH bonds present in the complexes, suggesting that the complexes were immobilized on the support.

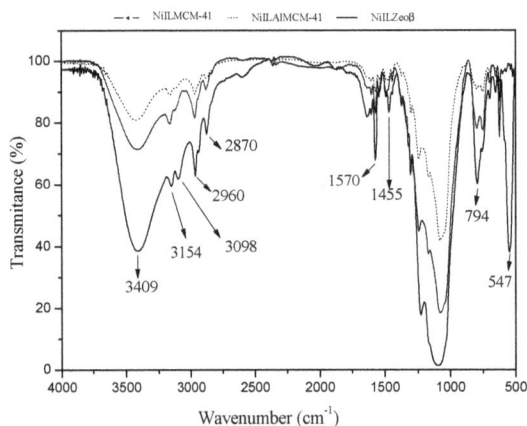

Figure 2. Infrared spectra of NiIL-Zeoβ, NiIL-MCM-41 and NiIL-Al-MCM-41.

Figure 3 shows the diffractograms that contain peaks characteristic of supports modified with an ionic liquid (IL-Zeoβ, IL-MCMC-41 and IL-Al-MCM-41) as well as β-zeolite, MCM-41 and Al-MCMC-41 (upper right insets). Based on a comparison of the sets of spectra, we conclude that the network organization did not change, which confirms that the structure of these supports did not changed even after the immobilization of the silylated ionic liquid in their cavities. The spectra in Figure 3 are consistent with the XRD patterns reported in the literature [19].

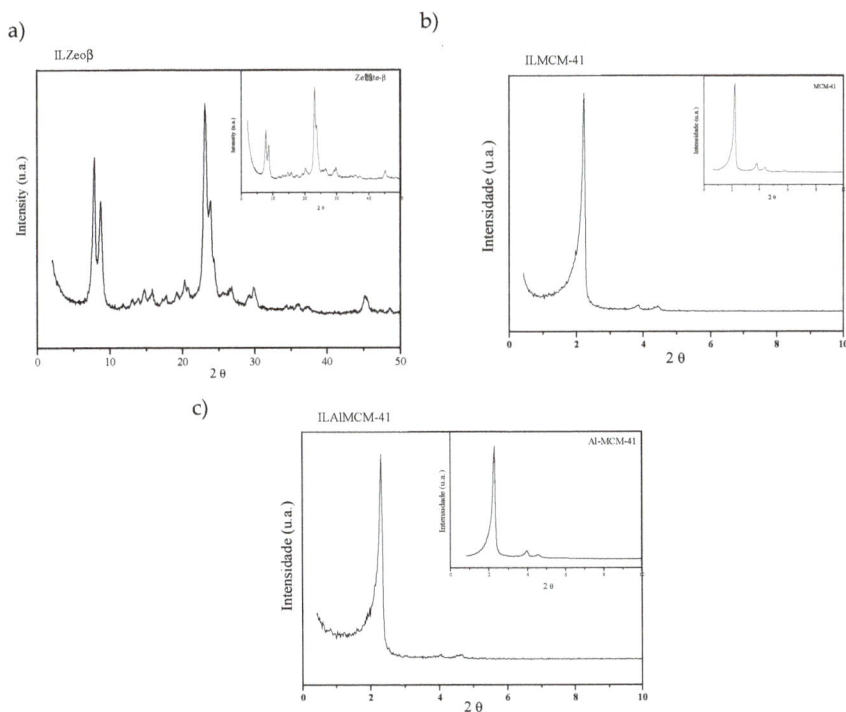

Figure 3. X-ray diffractograms of the (**a**) β-zeolite and IL-Zeoβ; (**b**) MCM-41 (Mobil Composition of Matter 41) and IL-MCM-41; and (**c**) Al-MCM-41 and IL-Al-MCM-41.

Evaluation of specific regions of the spectra can be used to confirm the results obtained by XRD wherein the characteristics of the materials were maintained, even after incorporation of the ionic liquid. Analysis of these data showed significant reductions in the specific surface areas of the IL-Zeoβ from 418 to 32 m^2 g^{-1}, the IL-MCM-41 from 1038 to 403 m^2 g^{-1} and the IL-Al-MCM-41 from 860 to 465 m^2 g^{-1} due to the incorporation of a substantial number of organic groups. These results suggest that the ionic liquid can be immobilized in the pores of the materials.

Based on the elemental analysis, one can estimate the amount of ionic liquid that was anchored to each support. We obtained 8.92% C, 1.39% H and 2.29% N on the IL-Zeoβ support; 7.28% C, 2.14% H and 2.31% N on the IL-MCM-41 support; and 10.57% C, 2.23% H and 2.76% N on the IL-Al-MCM-41 support. From these results, one can calculate the amount of ionic liquid incorporated into each medium; the relevant results were 22.0% (indicating that 41% of ionic liquid added was immobilized on the Zeoβ support), 22.2% (indicating that 47% of ionic liquid added was immobilized on IL-MCM-41) and 27.5% (indicating that 61% of ionic liquid added was immobilized on IL-Al-MCM-41).

The results of texture analysis from N_2 adsorption–desorption isotherms are shown in Table 2.

The specific areas were calculated by the BET method, resulting in 1038 m^2 g^{-1} for the MCM-41, 860 m^2 g^{-1} for the Al-MCM-41 and 418 m^2 g^{-1} for the β-zeolite, and 403, 465 and 32 m^2 g^{-1} for IL-MCM-41, IL-Al-MCM-41 and IL-β-zeolite, respectively. As expected, a decrease in BET surface area is observed in the ionic liquid-modified materials.

Table 2. Structural properties of the synthesized materials.

Support	A_{BET} [a] (m^2 g^{-1})	Pore Diameter [b] (Å)
MCM-41	1038	34
IL-MCM-41	403	28
Al-MCM-41	860	29
IL-Al-MCM-41	465	25
Zeoβ	418	-
IL-Zeoβ	32	-

[a] A_{BET} = specific area obtained using the BET method (total area). [b] Dp = average pore diameter calculated by intervals obtained using the BJH desorption method.

To confirm the binding of the nickel complex to the β-zeolite, MCM-41 and Al-MCM-41, the heterogenized complexes were analyzed by flame atomic absorption. Analyses were done in triplicate, and the values are reported in Table 3. The amount of nickel incorporated per gram of support is 75 ± 5% of added nickel, indicating the total immobilization of the complex. The NiIL-Zeoβ complex was also analyzed after an oligomerization reaction, and the results indicate that more than 50% of the immobilized complex in β-zeolite was lost during this reaction.

Table 3. The results from the atomic absorption spectroscopy (AAS) analysis of the β-diimine/β-zeolite nickel complex.

Support	(mmol Ni/g of Support)	(mmol Ni/g of Support)
	Added	by AAS
NiIL-Zeoβ		0.21
NiIL-MCM-41	0.3	0.23
NiIL-Al-MCM-41		0.24
NiIL-Zeoβ		0.10 [a]

[a] Amount of nickel measured by AAS after a reaction run.

3.2. Catalytic Tests Results

3.2.1. Homogeneous Catalytic Reactions

Homogeneous reactions were conducted to assess the selectivity and activity of the nickel-β-diimine complex so that those parameters could be compared with those of the catalytic tests performed with the heterogenized catalyst probably due to decomposition.

Table 4 presents the results of the catalytic tests. The ratio of Al/Ni was varied between 25 and 100 (entries 1–4). The highest activity was achieved at an Al/Ni ratio of 50, and the highest selectivity in butenes at an Al/Ni ratio of 25. The increase in activity with the Al/Ni ratio may be explained by an increase in the number of active species in the reaction medium. The maximum reactivity occurs before a decrease in activity, indicating catalyst deactivation, probably due to decomposition.

Table 4. Activity and selectivity in homogeneous catalytic reactions.

Entry	Catalyst (μmol)	Al/Ni	T (°C)	Pressure (atm)	TOF (s⁻¹)	Selectivity (%)		
						C$_4$ (1-C$_4$)	C$_6$	C$_{\geq 8}$
1	5	25	10	5	4.7 ± 0.1	85 (77)	10	5
2	5	50	10	5	19.3 ± 0.2	78 (45)	20	2
3	5	75	10	5	15.9 ± 0.2	77 (31)	17	6
4	5	100	10	5	14.4 ± 0.2	85 (48)	13	2
5	5	100	23	5	73.1 ± 0.2	79 (26)	19	2
6	10	50	10	5	15.7 ± 0.2	76 (31)	19	5
7	5	50	10	10	18.0 ± 0.2	80 (40)	17	3

Reaction conditions: solvent = toluene, cocatalyst = ethylaluminum sesquichloride (EASC).

When the amount of catalyst was changed from 5 to 10 μmol (entries 2 and 6), the activity and selectivity of the system decreased. This may be interpreted as resulting from an increase in the number of moles of catalyst precursor, which promotes an increase in the amount of active species in the reaction. Thus, the relative amount of ethylene reacting with the catalytic species is reduced, resulting in lower activity and selectivity in butenes. The reduction was mainly observed in 1-butene selectivity, due to parallel isomerization reaction. C$_6$ production is not significantly affected, and C$_8$ products slightly increase. This is consistent with the results reported by Busico et al. [29].

By increasing the initial reaction temperature from 10 to 23 °C (entries 4 and 5), the activity of the system also increases. Dubois et al. and Souza et al. [30,31] studied the influences of reaction temperature on oligomerization when using nickel complexes and found that the 1-butene best selectivity is achieved at low temperatures. Therefore, the subsequent reactions were performed at 10 °C.

The last parameter to be studied was the pressure (entries 2 and 7). In the determination of the effects of pressure on the activity and selectivity in ethylene oligomerization, it can be observed that there are no meaningful changes in the selectivity of the system, likely because the system is close to ethylene saturation.

Comparing these results with the industrial Shell High Olefins Process (SHOP), our oligomerization process is more selective to light olefins (C$_4$–C$_8$), while SHOP results in a larger range (C$_4$–C$_{40}$).

3.2.2. Catalytic Tests in Heterogeneous Reactions

The catalytic properties of the β-zeolite-supported nickel-β-diimine complex were tested under different reaction conditions, as shown in Table 5. These experiments were performed to find the conditions under which the reaction showed the best activity and selectivity. We used the same amount of heterogenized catalyst precursor as was determined in the optimization of the homogeneous catalysis (i.e., 5 μmol).

The first parameter studied was the Al/Ni molar ratio. Entries 8, 9 and 10 present the results of catalytic tests with different molar ratios of aluminum from the cocatalyst and nickel. These results show that as we increase the Al/Ni molar ratio, the activity of the catalyst also increases due to the increase in the concentration of active species present in the reaction mixture.

Table 5. Results of catalytic tests of the complex supported on β-zeolite modified with ionic liquid.

Entry	Al/Ni	T (°C)	Pressure (atm)	TOF (s⁻¹)	Selectivity (%)		
					C_4 (1-C_4)	C_6	$C_{\geq 8}$
8	25	10	5	1.1 ± 0.1	96 (94)	2	2
9	50	10	5	3.3 ± 0.2	96 (86)	4	-
10	100	10	5	3.8 ± 0.1	86 (74)	6	8
11	100	23	5	23.5 ± 1.0	100 (43)	-	-
12	50	10	2.5	3.0 ± 0.2	78 (89)	4	18.5
13	50	10	10	9.1 ± 0.2	94 (82)	4	2
14	50	10	5	5.0 ± 0.1	100 (91)	-	-
15	50	10	5	0.12 ± 0.01	100 (98)	-	-
16	50	10	5	0.06 ± 0.01	100 (98)	-	-

Reaction conditions: solvent = toluene, cocatalyst = EASC, $n_{catalyst}$ = 5 µmol.

After assessing the effect of the Al/Ni molar ratio, we set out to determine the optimal temperature for the reaction to obtain the best activity and selectivity in the production of butenes. As shown in entry 10, when the reaction was conducted at 10 °C, the selectivity was approximately 74% for 1-butene. Based on entry 11, when the reaction was performed at 23 °C, there was a decrease in the selectivity for 1-butene (approximately 43%) compared to entry 10. Using the conditions described above, this system shows higher selectivity than the homogeneous system (entries 4 and 5, Table 4), so the other reactions were performed at 10 °C.

The following studies were performed with pressures between 2.5, 5 and 10 atm to determine the effect of pressure on the activity of the oligomerization of ethylene. It can be observed from entries 9, 12 and 13 that the role of pressure in the system is very important. When comparing the results of reactions run at 2.5 and 5 atm, a slight difference in activity values can be observed, which can be explained by the difficulty of ethylene insertion into the cavities of zeolite due to the presence of the ionic liquid used as catalyst support. On the other hand, increasing the pressure from 5 to 10 atm increases the solubility of ethylene and, as a result, the ethylene enters the cavities, making it closer to the active species, which may explain the increased activity.

When comparing the homogeneous (entry 7, Table 4) and heterogeneous (entry 13, Table 5) reactions at 10 atm, it was observed that the heterogeneous reaction was more selective than the homogeneous reaction for 1-butene, which shows that the catalyst lies within the cavities of the β-zeolite and that the increase in selectivity is due to the confinement effect of the catalyst. However, based on the selectivities observed in hexenes and octenes, the reaction may also be occurring with the homogeneous catalyst, as demonstrated by atomic absorption analysis after the reaction of the nickel complex immobilized on β-zeolite. Our results show higher activity in heterogeneous media than is reported in the literature [7].

The next step was to test the catalytic properties of the NiIL-MCM-41 (entries 17, 18 and 19, Table 6) and NiIL-Al-MCM-41 (entry 20) catalysts under the best conditions obtained for the β-zeolite-supported β-diimine nickel complex: temperature (10 °C), moles of catalyst (5 µmol), pressure (10 atm) and reaction time (30 min). As shown in Table 5, these catalysts are less active than the β-zeolite-supported system; however, these catalysts are 100% selective for butenes. When the Al/Ni ratio was increased from 25 to 100, it was found that, at an Al/Ni ratio of 50, the selectivity of the system did not vary significantly with respect to the activity; the catalytic system was more active, which indicates an increase in the active species present in the reaction medium.

Table 6. Results of catalytic tests of the complex supported on MCM-41 and Al-MCM-41 modified with ionic liquid.

Entry	Al/Ni	TOF (s^{-1})	Selectivity (%)		
			C$_4$ (1-C$_4$)	C$_6$	C$_{\geq 8}$
17	25	1.0 ± 0.1	100 (94)	-	-
18	50	1.4 ± 0.1	100 (96)	-	-
19	100	1.0 ± 0.1	100 (93)	-	-
20 *	100	2.3 ± 0.1	100 (90)	-	-

* NiILAl-MCM-41. Reactive conditions: solvent = toluene, cocatalyst = EASC, $n_{catalyst}$ = 5 μmol, temperature = 10 °C, pressure = 10 atm.

In our analysis of the results of the catalytic test with NiIL-Al-MCM-41, we observed that the system was highly active (approximately 2.3 s^{-1}) and reached 90% selectivity for 1-butene, which makes its activity superior to that of the NiIL-MCM-41 system.

Comparing the selectivities of the three heterogeneous and homogeneous systems, we observed that the effect of the confinement of the catalyst inside the MCM-41 pores was a determining factor in the high selectivity of the heterogeneous system, in this case. However, when comparing the supports used, we observed that the difference in the pore size of these materials was not a determining factor for the activity of the reactions, but rather the presence of acid sites in the structure of the β-zeolite and Al-MCM-41 was a determining factor. Hulea et al. [32] reported that the content of both nickel and acid sites contribute to the activation of the oligomerization reaction.

4. Conclusions

In this work, we described synthesis and caractherization of nickel-β-diimine complexes immobilized on β-zeolite, [Si]-MCM-41 and [Si,Al]-MCM-41 modified with an ionic liquid. These catalytic systems were used in ethylene oligomerization reactions and compared with the same reactions in homogeneous media. Catalytic properties of synthesized materials were tested under different conditions, varying Al/Ni molar ratio, temperature, pressure and catalyst loading.

The results obtained in oligomerization reactions demonstrate the efficiency of heterogenized catalystic system. The techniques of characterization of the different complexes supported on β-zeolite or mesoporous materials of the MCM type corroborate that immobilization of the complex did not alter the structure of the supports of Si and Al.

Comparison between the heterogeneous systems indicates that differences in pore sizes of the support materials did not influence the activity, but the presence and strength of acid sites, occurring in β-zeolite and Al-MCM-41, affects reaction activity. We suggest that the effect of acid sites of the support on the metal site is to increase the strength of ethylene coordination to the metal, causing the system to be more active.

Comparison between homogeneous and heterogeneous systems containing the β-diimine-nickel complex showed that heterogenized systems are more selective in the formation of 1-butene. This higher selectivity indicates a positive relationship between porosity and confinement of the catalyst within channels of the support materials.

Author Contributions: Katia Bernardo-Gusmão and Roberto Fernando de Souza conceived and designed the experiments; Camila A. Busatta performed the experiments; Katia Bernardo-Gusmão, Camila Aguilar Busatta and Marcelo Luis Mignoni analyzed the data and wrote the paper.

Acknowledgments: The authors thank CNPq for their financial support in the form of grants.

Conflicts of Interest: The authors declare no conflict of interest.

References

1. Freitas, E.R.; Gum, C.R. Shells higher olefins process. *Chem. Eng. Prog.* **1979**, *25*, 73–76.
2. Chauvin, Y.; Gaillard, J.F.; Quang, D.V.; Andrews, J.W. IFP Dimersol process for dimerization of C3 and C4 olefinic cuts. *Chem. Ind.* **1974**, 375–378.
3. Commereuc, D.; Chauvin, Y.; Gaillard, J.F.; Léonard, J.; Andrews, J. Dimerize ethylene to butene-1. *Hidrocarbon Process* **1984**, *63*, 118–120.
4. Rossetto, E.; Nicola, B.P.; de Souza, R.F.; Pergher, S.B.C.; Bernardo-Gusmão, K. Anchoring via covalent binding of beta-diimine-nickel complexes in SBA-15 and its application in catalytic reactions. *Appl. Catal. A* **2015**, *502*, 221–229. [CrossRef]
5. Rossetto, E.; Caovilla, M.; Thiele, D.; de Souza, R.F.; Bernardo-Gusmão, K. Ethylene oligomerization using nickel-beta-diimine hybrid xerogels produced by the sol-gel process. *Appl. Catal. A* **2013**, *454*, 152–159. [CrossRef]
6. De Souza, M.O.; de Souza, R.F.; Rodrigues, L.R.; Pastore, H.O.; Gauvin, R.M.; Gallo, J.M.R.; Favero, C. Heterogenized nickel catalysts for propene dimerization: Support effects on activity and selectivity. *Catal. Commun.* **2013**, *32*, 32–35. [CrossRef]
7. Rossetto, E.; Nicola, B.P.; de Souza, R.F.; Bernardo-Gusmão, K.; Pergher, S.B.C. Heterogeneous complexes of nickel MCM-41 with beta-diimine ligands: Applications in olefin oligomerization. *J. Catal.* **2015**, *323*, 45–54. [CrossRef]
8. Martins, L.; Cardoso, D. Catalytic applications of basic micro and mesoporous molecular sieves. *Quím. Nova* **2006**, *29*, 358–364. [CrossRef]
9. Beck, J.S.; Vartuli, J.C.; Roth, W.J. A new family of mesoporous molecular sieves prepared with liquid-cristal. *J. Am. Chem. Soc.* **1992**, *114*, 10834–10843. [CrossRef]
10. Guo, C.Y.; Xu, H.; Zhang, M.; Zhang, X.; Yan, F.; Yuan, G. Immobilization of *bis*(imino)pyridine iron complexes onto mesoporous molecular sieves and their catalytic performance in ethylene oligomerization. *Catal. Commun.* **2009**, *10*, 1467–1471. [CrossRef]
11. Van Looveren, L.K.; de Vos, D.E.; Vercruysse, K.A.; Geysen, D.F.; Janssen, B.; Jacobs, P.A. Oligomerization of propene on an alumoxane-grafted MCM-41 host with *bis*(cyclopentadienyl)zirconium dimethyl $(Cp_2Zr(CH_3)_2)$. *Catal. Lett.* **1998**, *56*, 53–56. [CrossRef]
12. Shao, H.; Zhou, H.; Guo, X.; Tao, Y.; Jiang, T.; Qin, M. Chromium catalysts supported on mesoporous silica for ethylene tetramerization: Effect of the porous structure of the supports. *Catal. Commun.* **2015**, *60*, 14–18. [CrossRef]
13. De Souza, M.O.; Rodrigues, L.R.; Pastore, H.O.; Ruiz, J.A.C.; Gengembre, L.; Gauvin, R.M.; de Souza, R.F. A nano-organized ethylene oligomerization catalyst: Characterization and reactivity of the $Ni(MeCN)_6(BF_4)_2/[Al]$-MCM-41/$AlEt_3$ system. *Microporous Mesoporous Mater.* **2006**, *96*, 109–114. [CrossRef]
14. De Souza, M.O.; Rodrigues, L.R.; Gauvin, R.M.; de Souza, R.F.; Pastore, H.O.; Gengembre, L.; Ruiz, J.A.C.; Gallo, J.M.R.; Milanesi, T.S.; Milani, M.A. Support effect in ethylene oligomerization mediated by heterogenized nickel catalysts. *Catal. Commun.* **2010**, *11*, 597–600. [CrossRef]
15. Borba, K.M.N.; de Souza, M.O.; de Souza, R.F.; Bernardo-Gusmão, K. beta-Diimine nickel complexes in BMI center dot $AlCl_4$ ionic liquid: A catalytic biphasic system for propylene oligomerization. *Appl. Catal. A* **2017**, *538*, 51–58. [CrossRef]
16. Thiele, D.; de Souza, R.F. Biphasic ethylene oligomerization using *bis*(imino)pyridine cobalt complexes in methyl-butylimidazolium organochloroaluminate ionic liquids. *J. Mol. Catal. A* **2011**, *340*, 83–88. [CrossRef]
17. Mehnert, C.P.; Cook, R.A.; Dispenziere, N.C.; Afeworki, M. Supported ionic liquid catalysis—A new concept for homogeneous hydroformylation catalysis. *J. Am. Chem. Soc.* **2002**, *124*, 12932–12933. [CrossRef] [PubMed]
18. Riisager, A.; Wasserscheid, P.; van Hal, R.; Fehrmann, R. Continuous fixed-bed gas-phase hydroformylation using supported ionic liquid-phase (SILP) Rh catalysts. *J. Catal.* **2003**, *219*, 452–455. [CrossRef]
19. Panda, A.G.; Bhor, M.D.; Jagtap, S.R.; Bhanage, B.M. Selective hydroformylation of unsaturated esters using a Rh/PPh_3-supported ionic liquid-phase catalyst, followed by a novel route to pyrazolin-5-ones. *Appl. Catal. A* **2008**, *347*, 142–147. [CrossRef]
20. Riisager, A.; Jørgensen, B.; Wasserscheid, P.; Fehrmann, R. First application of supported ionic liquid phase (SILP) catalysis for continuous methanol carbonylation. *Chem. Commun.* **2006**, 994–996. [CrossRef] [PubMed]

21. Mehnert, C.P.; Mozeleski, E.J.; Cook, R.A. Supported ionic liquid catalysis investigated for hydrogenation reactions. *Chem. Commun.* **2002**, 3010–3011. [CrossRef]

22. Hagiwara, H.; Sugawara, Y.; Isobe, K.; Hoshi, T.; Suzuki, T. Immobilization of $Pd(OAc)_2$ in ionic liquid on silica: Application to sustainable Mizoroki-Heck reaction. *Org. Lett.* **2004**, *6*, 2325–2328. [CrossRef] [PubMed]

23. Yamaguchi, K.; Yoshida, C.; Uchida, S.; Mizuno, N. Peroxotungstate immobilized on ionic liquid-modified silica as a heterogeneous epoxidation catalyst with hydrogen peroxide. *J. Am. Chem. Soc.* **2005**, *127*, 530–531. [CrossRef] [PubMed]

24. Tang, L.; Duan, Y.; Li, X.; Li, Y. Syntheses, structure and ethylene polymerization behavior of beta-diiminato titanium complexes. *J. Organomet. Chem.* **2006**, *691*, 2023–2030. [CrossRef]

25. Feldman, J.; McLain, S.J.; Parthasarathy, A.; Marshall, W.J.; Calabrese, J.C.; Arthur, S.D. Electrophilic metal precursors and a beta-diimine ligand for nickel(II)- and palladium(II)-catalyzed ethylene polymerizations. *Organometallics* **1997**, *16*, 1514–1516. [CrossRef]

26. Sahoo, S.; Kumar, P.; Lefebvre, F.; Halligudi, S.B. Oxidative kinetic resolution of alcohols using chiral Mn-salen complex immobilized onto ionic liquid modified silica. *Appl. Catal. A* **2009**, *354*, 17–25. [CrossRef]

27. Mignoni, M.L.; de Souza, M.O.; Pergher, S.B.C.; de Souza, R.F.; Bernardo-Gusmão, K. Nickel oligomerization catalysts heterogenized on zeolites obtained using ionic liquids as templates. *Appl. Catal. A* **2010**, *374*, 26–30. [CrossRef]

28. Corma, A.; Kan, Q.; Navarro, M.T.; Pariente, J.P.; Rey, F. Synthesis of MCM-41 with different pore diameters without addition of auxiliary organics. *Chem. Mater.* **1997**, *9*, 2123–2126. [CrossRef]

29. Busico, V.; Cipullo, R. Influence of monomer concentration on the stereospecificity of 1-alkene polymerization promoted by C-2-simmetrical ansa-metallocene catalysts. *J. Am. Chem. Soc.* **1994**, *116*, 9329–9330. [CrossRef]

30. Sauthier, M.; Leca, F.; de Souza, R.F.; Bernardo-Gusmão, K.; Queiroz, L.F.T.; Toupet, L.; Réau, R. $NiCl_2$(1,2-diiminophosphorane) complexes: A new family of readily accessible and tuneable catalysts for oligomerisation of ethylene. *New J. Chem.* **2002**, *26*, 630–635. [CrossRef]

31. De Souza, R.F.; Bernardo-Gusmão, K.; Cunha, G.A.; Loup, C.; Leca, F.; Réau, R. Ethylene dimerization into 1-butene using 2-pyridylphosphole nickel catalysts. *J. Catal.* **2004**, *226*, 235–239. [CrossRef]

32. Hulea, V.; Fajula, F. Ni-exchanged AlMCM-41—An efficient bifunctional catalyst for ethylene oligomerization. *J. Catal.* **2004**, *225*, 213–222. [CrossRef]

© 2018 by the authors. Licensee MDPI, Basel, Switzerland. This article is an open access article distributed under the terms and conditions of the Creative Commons Attribution (CC BY) license (http://creativecommons.org/licenses/by/4.0/).

applied
sciences

MDPI

Article

Mesoporous Niobium Oxyhydroxide Catalysts for Cyclohexene Epoxidation Reactions

Izabela D. Padula, Poliane Chagas, Carolina G. Furst and Luiz C. A. Oliveira *

Departamento de Química, Universidade Federal de Minas Gerais, Belo Horizonte 31270-901, Brazil;
izabelapadula@gmail.com (I.D.P.); polianechagas@gmail.com (P.C.); carolfurstg@gmail.com (C.G.F.)
* Correspondence: luizoliveira@qui.ufmg.br

Received: 18 April 2018; Accepted: 18 May 2018; Published: 28 May 2018

Abstract: Mesoporous catalysts were synthesized from the precursor $NbCl_5$ and surfactant CTAB (cetyltrimethylammonium bromide), using different synthesis routes of, in order to obtain materials with different properties which are capable of promoting the epoxidation of cyclohexene. The materials were characterized by X ray diffractometry (XRD), thermogravimetry (TG), acidity via pyridine adsorption, Hammet titration and N_2 adsorption/desorption. The characterization data indicate that the calcination process of the catalysts was efficient for elimination of the surfactant, but it caused a collapse of the structure, causing a Brunauer Emmett Teller (BET) specific area decrease (ClNbS-600, 44 and ClNbS-AC-600, 64 m^2 g^{-1}). The catalysts that have not been calcined showed high BET specific areas (ClNbS 198 and ClNbS-AC 153 m^2 g^{-1}). Catalytic studies have shown that mild reaction conditions promote high conversion. The catalysts ClNbS and ClNbS-AC showed high conversions of cyclohexene, 50 and 84%, respectively, while the calcined materials showed low conversion (<30%). The epoxide formation was confirmed by nuclear magnetic resonance (NMR).

Keywords: epoxidation; cyclohexene; niobium oxyhydroxide catalysts

1. Introduction

The use of molecular sieves has been of great industrial interest, since they direct the reactions due to their defined porous structure, being able to promote a greater selectivity for the product of interest. Two classes of well-defined and widely studied porous materials are the microporous zeolites [1] and the mesoporous silicates known as Mobil Composition of Mater (MCM) [2]. Moreover, the incorporation of transition metals, such as those of group V, in these porous materials can generate materials with high catalytic activity and good selectivity [1]. The presence of mesopores allows compounds of higher kinetic volume to access the active sites deposited in the cavities, which is difficult when using supports with smaller pores, such as the microporous. Niobium compounds stand out among those of neighboring elements because of their bifunctional properties, since they may have high acidity and ability to form highly oxidizing species in the presence of H_2O_2 by the formation of peroxo (Nb-O-OH) surface groups [3]. This allows their oxides to be widely used in heterogeneous catalysis in several types of reactions [4]. In addition, the bifunctional character of niobium oxides has been explored in reactions that require dehydration/oxidation properties of the substrate [5]. The niobium oxides commonly used have low specific area and low mesoporosity [6], which diminishes their activity as catalysts in reactions of total or partial oxidation involving molecules with higher kinetic volume. Selective oxidation reactions are less thermodynamically favored than those that promote total oxidation, necessitating a suitable catalyst to direct the formation of the product of kinetic interest [7].

Epoxidation of olefins is a selective oxidation reaction, especially in the production of various raw materials, as well as in the synthesis of intermediates for the production of chemical and pharmaceutical products [8–11]. Epoxidation may occur via homogeneous catalysis, but the greatest disadvantage of this process is the recovery of the catalyst, increasing interest in the development of heterogeneous catalysts capable of performing this reaction [12]. In the literature there are several studies on the epoxidation of olefin model molecules, and V and Nb compounds have been widely used but supported on porous materials. Using porous materials Nogueira et al., obtained near 100% conversion of the toluene oxidation reactions [13]. Rohit et al., employing vanadium-titania catalysts obtained conversion of 94% for the epoxidation of a variety of alkenes with organic solvent extracted TBHP as the oxidant [14]. In another work, Gallo et al. employing catalysts based on niobium metallocenes deposited into mesoporous silica showed conversion of 58% and high selectivity to limonene epoxide >98% [15]. Works using niobium compounds are scarce, mainly dealing with synthetic compounds where their textural properties can be modulated, in order to obtain high activity in epoxidation reactions.

Thus, it is understood that it is of great technological interest to obtain catalysts based on niobium compounds that can be efficiently used in selective oxidation reactions of olefins, aiming at the production of compounds of high added value that are of interest to the petrochemical industry. Different niobium catalyst synthesis methods are mentioned in the literature [16–20] as well as studies on the activity of these materials [21]. Therefore, we propose the synthesis of niobium oxyhydroxide from $NbCl_5$ using a surfactant to increase the hydrophobicity of the catalysts in order to improve their interaction with the non-polar substrate, and we test these materials for their ability to realize the epoxidation of cyclohexene used as a model molecule.

2. Materials and Methods

2.1. Synthesis of Catalysts

The synthesis of ClNbS and ClNbS-600 catalysts was performed by dissolving 41.2 mmols of CTAB in 25 mL of 1-butanol (99% VETEC), 69 mL of 1-hexanol (99% VETEC) and 19 mL of deionized water in a beaker at 60 °C with constant stirring. After dissolution, 46.2 mmol of NbCl5 were added. After the total solubilization of the salt, NH_4OH (5 mol L^{-1}) was dripped to pH 7. The volume of the formed solution was completed to 1000 mL with deionized water at 60 °C and the system was left under constant magnetic stirring for two (2) days for aging to yield the hydrophobized niobium oxyhydroxide. The solid formed was then macerated in an agate mortar. Part of the macerated solid was subjected to a heat treatment, remaining at 600 °C for 3 h, following a heating ramp of 10 °C min^{-1}. The catalysts ClNbS and ClNbS-600, respectively, were thus obtained.

The ClNbS-AC and ClNbS-AC-600 catalysts were synthesized as described above, however after pH correction the formed solution was poured into a Bergof ® BR100 stainless steel autoclave containing one teflon beaker; then 20 mL of deionized water were added. The system was left at 60 °C under constant magnetic stirring for 2 (two) days for aging to produce the hydrophobized niobium oxyhydroxide. The formed product was transferred to a beaker with 100 mL of deionized water and a white solid suspension formed on top. The suspension was washed with deionized water and oven dried at 60 °C for 12 h. The ClNbS-AC-600 catalyst underwent the same calcination process as ClNbS-600.

2.2. Catalysts Characterization

The textural properties of the materials were analyzed through a N_2 adsorption/desorption isotherm that was obtained at a temperature of 77 K using a Quantachrome Autosorb IQ_2 equipment and pore volumes were calculated based on the adsorption branch of the isotherm using the Barrett Joyner Halenda (BJH) method. The specific surface area value was obtained using the BET method. X ray diffraction analysis was performed using the SHIMADZU model XRD-7000 X ray diffractometer, equipped with a copper tube and graphite monochromator. Analyses were performed under 30 mA current and 30 kV voltage. The velocity used was 4 degrees min^{-1} for scanning between the angles

$10° < 2\theta < 70°$ by applying time constants of 5 s per increment. Infrared spectra (FTIR) were obtained using a Perkin Elmer FTIR RXI spectrophotometer in the region of 4000 to 500 cm^{-1} with a resolution of 4 cm^{-1} and a mean scan of the signal equal to 64 scans. The thermal analyzes were performed on a Shimadzu-TGA50H thermocouple, up to a maximum temperature of 700 °C, in air atmosphere, under a flow of 100 mL min^{-1} at a heating rate of 10 °C min^{-1}, using approximately 3 mg of sample. The acidic properties of the catalyst were determined by the pyridine adsorption method. In this test, 10 mg of catalyst was placed in crucibles inside a quartz tube in a furnace and heated to 100 °C, for 2 h, under air flow (80 mL min^{-1}). After the cleaning of the surface, the pyridine was then introduced, at 50 °C. After the adsorption step, the temperature was raised at 100 °C under air flow (80 mL min^{-1}) to remove the pyridine physisorbed on the catalyst surface. Discs of 1 cm of diameter were made under vacuum and 6 ton cm^{-2}. The infrared spectrum was acquired using a Spectrm RX, 64 scans were recorded, in a region of 1800–1400 cm^{-1}. Quantitative determination of the acid sites present in all catalysts was performed by Hammett titration as described by Zhao et al. [22].

2.3. Catalytic Tests

Cyclohexene epoxidation reactions were performed in the presence of cyclohexene (VETEC, 99%) 50% hydrogen peroxide (SYNTH) and acetonitrile (J.T. BAKER, 99%).

Batch type glass reactors with a capacity of 15 mL and autogenous pressure were used. The reuse for the best catalyst was studied.

The initial condition studied was adapted according to experiments indicated in the literature [23] using 20 mmol of cyclohexene, 2 mL of 50% hydrogen peroxide, 50 mg of catalyst in 10 mL of acetonitrile, at 25 °C for 1 h and constant stirring.

After the completion of each reaction, the catalyst was separated by centrifugation. Sample preparation consisted of the addition of 20 μL of heptane (VETEC, P.A) as internal standard (PI) in 0.98 mL of sample. They were then analyzed by gas chromatography with flame ionization detector (GC-FID), GC-Shimadzu, DB-5 ((5% -phenyl) -methylpolysiloxane column), 30 m × 0.32 mm × 0.25 μm. The parameters used were column temperature: 40 °C for 10 min; injection temperature: 200 °C with Split (1:25); injected volume: 0.4 μL; FID detector temperature: 220 °C; and entrainment gas N$_2$ with flow of 0.6 mL min^{-1}. The CLASS CR-10 software was used for data acquisition. The conversion was determined from a calibration curve with internal standard.

To determine the formation of the reaction products, solutions containing 300 μL of reaction were prepared in 300 μL of D$_2$O. These solutions were analyzed in a Bruker AVANCE DPX 200 Nuclear Magnetic Resonance Spectrometer. ^{1}H and ^{13}C spectra were acquired, in which 16 and 128 scans were used, respectively. Analyzes were performed for the calibration curve, in which the preparation of the sample used the CDCl$_3$ as solvent.

3. Results and Discussion

3.1. Characterization of Catalysts

The results obtained through thermogravimetric analysis (Figure 1a) show that materials that were not calcined (ClNbS and ClNbS-AC) present three main mass loss events. The first two must be related to external and bulk hydroxylates of the catalysts. The mass loss above 350 °C is attributed to the decomposition of the surfactant group anchored to the surface of the material [24]. No significant mass loss events were observed for calcined catalysts (ClNbS-600 and ClNbS-AC-600); which means that the calcination temperature was sufficient to remove all surfactant and hydroxyls, leaving only Nb$_2$O$_5$. The X ray diffractograms (Figure 1b) showed higher crystallinity for the calcined catalysts, since those that were not thermally treated presented as totally amorphous. The catalysts ClNbS-600 and ClNbS-AC-600 presented a typical diffraction profile of niobium oxide (Nb$_2$O$_5$), consistent with the standard JCPDS 27-1003.

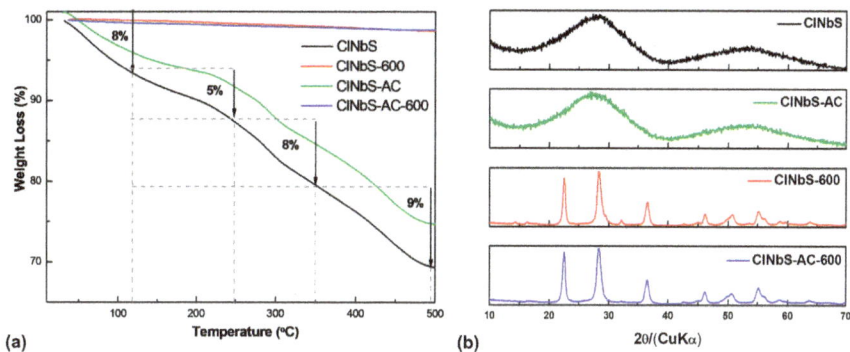

Figure 1. (**a**) Thermogravimetric curves for the catalysts under air atmosphere and (**b**) X ray diffraction patterns of ClNbS-600 and ClNbS-AC-600 catalysts.

The N_2 adsorption/desorption analyzes were performed and are presented in Figure 2. The data presented in Table 1 show that the calcination process decreases the specific area in relation to non-calcined catalysts, is attributed to the occurrence of particle agglomeration and crystallization, reducing the likely porous structure of the material. Another relationship observed in the surface area data is in relation to the form of aging: the ClNbS catalyst presents a larger surface area than ClNbS-AC, which may be associated with larger organization of the ClNbS-AC pores, since there was a control pressure and temperature in the synthesis.

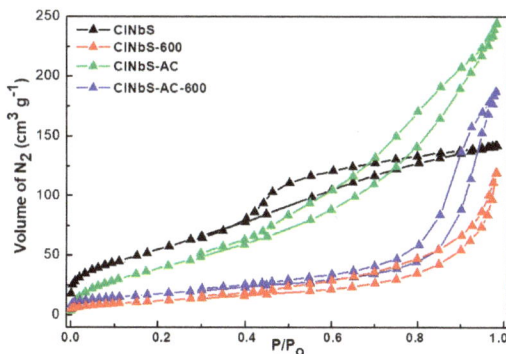

Figure 2. N_2 adsorption/desorption of the catalysts.

Table 1. BET specific area values and pore volume by the BJH method for the synthesized catalysts.

Catalyst	BET Surface Area ($m^2 g^{-1}$)	Pore Volume ($cm^3 g^{-1}$)
ClNbS	198	0.17
ClNbS-600	44	0.10
ClNbS-AC	153	0.15
ClNbS-AC-600	64	0.09

The isotherms of the catalysts ClNbS-AC and ClNbS-600 indicate the absence of pores, being type II, according to the classification suggested by IUPAC, typical of non-porous materials. The hysteresis formed by the desorption isotherm of these catalysts can be classified as H3, as it does not reach a saturation level and is reported for aggregate materials, forming pores of the crack type associated with a portion of interparticle pores [25,26].

In the case of the ClNbS and ClNbS-AC-600 samples, the profile of the type IV isotherm present in the same proportion of materials suggests the presence of mesopores, confirmed by the pore size distribution (not shown here). When reaching a saturation level in the adsorption of $N_{2(g)}$, during the desorption, the ClNbS forms a hysteresis classified as H2, considered as bottle-type pores [26]. For the ClNbS-AC-600 the formed hysteresis is classified as H3. Table 1 shows the BET specific area values and pore volume of the materials.

The calcined catalysts were subjected to acidity analysis by adsorption of pyridine to qualitatively evaluate the type of acid sites present in these materials. For the non-calcined catalysts the analyzes were not carried out because the surfactant had adsorption bands in the same region where the pyridine adsorbed at the Lewis sites (1447 cm^{-1}) and adsorbed at the Bronsted sites (1550 cm^{-1}) were observed. The ClNbS-600 catalyst exhibits acidity Bronsted, while ClNbS-AC-600 catalyst presented Bronsted acid and Lewis acid sites (Figure 3), indicating that the aging process in an autoclave provides a difference in the type of acidic material. It is important to consider the acidity of the materials, since it is believed that the oxidant groups are formed in these acid sites by the decomposition of H_2O_2.

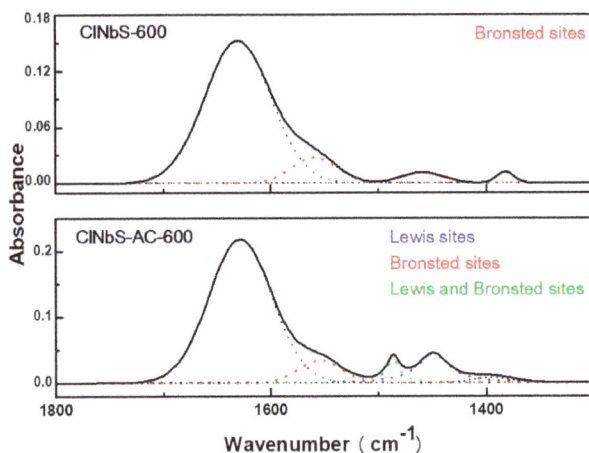

Figure 3. IR spectra for acidity analysis by pyridine adsorption on ClNbS-600 and ClNbS-AC-600 catalysts.

By Hammet titration the total amount of acid sites of the catalysts were determined and the mean acidity values were 2.02, 2.78, 1.80 and 2.60 mmol g^{-1} for ClNbS, ClNbS-600, ClNbS-AC and ClNbS-AC-600, respectively. The calcined catalysts had a higher amount of acid sites, possibly due to unclogging of the pores after surfactant elimination. The catalysts that passed through the aging process in the autoclave presented a smaller amount of acid sites, because this treatment caused a better organization of the catalysts structure, reducing the imperfections and faults on the surface of the catalysts, where possibly acidic sites would reside.

3.2. Epoxidation Reactions

Figure 4a shows the preliminary catalytic results for the catalysts before and after calcination. A first analysis identifies that the materials that did not undergo calcination presented a greater cyclohexene conversion capacity.

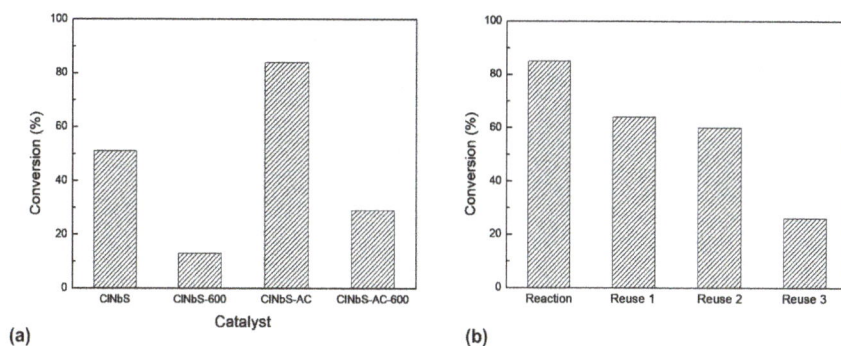

Figure 4. (**a**) Epoxidation reactions using niobium catalysts. Conditions: 25 °C; 60 min; molar ratio (2:1) H_2O_2/cyclohexene; 50 mg of catalyst. (**b**) ClNbS-AC catalyst reuse test at 25 °C, 60 min, 50 mg catalyst and H_2O_2/cyclohexene molar ratio 2:1.

The catalysts ClNbS-AC and ClNbS are niobium oxyhydroxides (NbO_2OH) and as proposed by Souza et al., these materials are capable of forming peroxo groups in the presence of H_2O_2. These groups are highly oxidizing, which would explain the better catalytic efficiency of the ClNbS-AC (80%) and ClNbS (50%) catalysts in the oxidation of cyclohexene [27]. The calcined catalysts do not present these surface hydroxyls for this reason they are not able to form peroxo groups, so their catalytic activity in the oxidation of cyclohexene are lower. The ClNbS-AC catalyst showed the highest cyclohexene conversion (84%) under the conditions studied. Ziolek et al., also analyzed the conversion of cyclohexene using niobium compounds as a catalyst under similar reaction conditions and obtained a maximum conversion of 58% [28].

The results of the catalyst reuse tests are presented in Figure 4b. A small conversion decrease in the cycles of reaction was observed, which may be related to loss of catalyst mass during the washes and the deactivation of the catalytic sites after some reactions, however the catalyst showed a certain stability, with considerable cyclohexene conversions (~60%) after two reuse tests. However, in the third reuse of the catalyst there is a marked drop in the conversion capacity. The catalyst was recovered after this test and it was observed that the hydrophobizing group had been leached from the surface after the third use. The presence of the group that makes the catalyst have greater affinity for the solvent where the substrate is dissolved [23]. Other forms of synthesis are being studied in order to maintain a more stable structure and increase its reuse capacity.

^{1}H and ^{13}C spectra were obtained to prove epoxide formation by comparing the ^{13}C chemical shifts of the experimental and simulated. The values of the chemical displacements are presented in Table 2.

Table 2. Experimental and simulated chemical displacement values for epoxide carbon.

	Experimental Chemical Shifts (ppm)	Simulated Chemical Shifts (ppm)
C1	51.22	52.07
C2	25.36	24.50
C3	20.53	19.47

The ^{13}C spectrum is shown in Figure 5. The presence of the acetonitrile solvent, the cyclohexene substrate and the major product, epoxide, are observed. Other cyclohexene oxidation products, such as cyclohexanone, were not formed as no ketone carbonyl signal was observed.

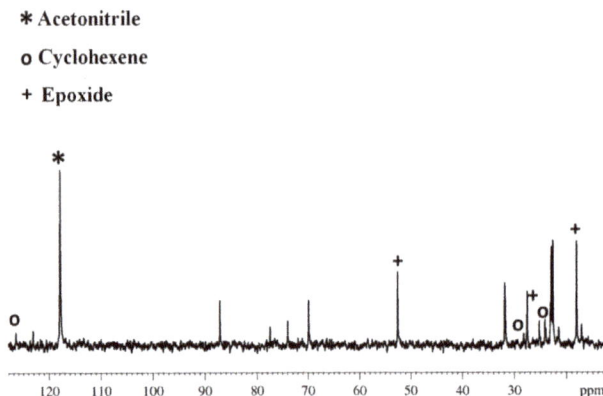

Figure 5. ^{13}C spectrum of the cyclohexene oxidation product using the ClNbS-AC catalyst. Spectrum acquired at room temperature on a 400 MHz spectrometer.

4. Conclusions

The results presented an efficient catalytic process using niobium oxyhydroxide synthesized from the precursor $NbCl_5$. The TG analysis showed the incorporation of CTAB in the structure of the material, as well as the loss of this group in the calcination process. The ClNbS and ClNbS-AC materials presented high BET specific area, with values of 198 and 153 m^2 g^{-1}, respectively. The calcination process resulted in a decrease of specific area and an increase in the acidity of the material due to the acid site increase and greater crystallinity presented for these materials.

The use of niobium oxyhydroxide with mesoporosity is rarely described in the scientific literature. The acidity generated by the hydroxyls can be replaced by oxidizing groups, i.e., peroxisome species which can be formed with compounds of the V group in the presence of hydrogen peroxide. The mesoporosity and formation of these oxidizing groups appears to provide important properties in partial oxidation reactions, such as epoxidation of olefins. The epoxidation reactions using cyclohexene, hydrogen peroxide showed good conversion results for the reaction conditions studied, especially ClNbS-AC, which presented 84% conversion. The NMR results showed the formation of the epoxide as the major product of the reaction. The results were extremely promising and innovative, given the efficiency of the catalysts in the reactions, as well as the fact that there have been no high conversion value reported in the literature for mild reaction conditions.

Author Contributions: I.D.P. and P.C. synthesized and characterized the catalysts used in this work. C.G.F and I.D.P performed the GC-FID analysis. All authors contributed to the writing, and analysis of this paper.

Conflicts of Interest: The authors declare no conflict of interest.

References

1. Marin-Astorga, N.; Martinez, J.J.; Borda, G.; Cubillos, J.; Suarez, D.N.; Rojas, H. Control of the chemoselectivity in the oxidation of geraniol over lanthanum, titanium and niobium catalysts supported on mesoporous silica MCM-41. *Top. Catal.* **2012**, *55*, 620–624. [CrossRef]
2. Yang, G.; Deng, Y.; Wang, J. Non-hydrothermal synthesis and characterization of MCM-41 mesoporous materials from iron ore tailing. *Ceram. Int.* **2014**, *40*, 7401–7406. [CrossRef]

3. De Oliveira, L.C.A.; Costa, N.T.; Pliego, J.R.; Silva, A.C.; de Souza, P.P.; Patrícia, P.S. Amphiphilic niobium oxyhydroxide as a hybrid catalyst for sulfur removal from fuel in a biphasic system. *Appl. Catal. B Environ.* **2014**, *147*, 43–48. [CrossRef]

4. Tanabe, K.; Okazaki, S. Various reactions catalyzed by niobium compounds and materials. *Appl. Catal. A Gen.* **1995**, *133*, 191–218. [CrossRef]

5. Mohd Ekhsan, J.; Lee, S.L.; Nur, H. Niobium oxide and phosphoric acid impregnated silica-titania as oxidative-acidic bifunctional catalyst. *Appl. Catal. A Gen.* **2014**, *471*, 142–148. [CrossRef]

6. Chan, X.; Pu, T.; Chen, X.; James, A.; Lee, J.; Parise, J.B.; Kim, D.H.; Kim, T. Effect of niobium oxide phase on the furfuryl alcohol dehydration. *Catal. Commun.* **2017**, *97*, 65–69. [CrossRef]

7. Alexopoulos, K.; Reyniers, M.F.; Marin, G.B. Reaction path analysis of propene selective oxidation over V_2O_5 and V_2O_5/TiO_2. *J. Catal.* **2012**, *295*, 195–206. [CrossRef]

8. Ramanathan, A.; Zhu, H.; Maheswari, R.; Thapa, P.S.; Subramaniam, B. Comparative study of Nb-incorporated cubic mesoporous silicates as epoxidation catalysts. *Ind. Eng. Chem. Res.* **2015**, *54*, 4236–4242. [CrossRef]

9. Santa A., A.M.; Vergara G., J.C.; Palacio S., L.A.; Echavarría I., A. Limonene epoxidation by molecular sieves zincophosphates and zincochromates. *Catal. Today* **2008**, *133–135*, 80–86. [CrossRef]

10. Moreira, M.A. *Produção Enzimática de Peróxi-Ácidos e sua Utilização na Epoxidação de Terpenos*; Universidade Federal de Santa Catarina: Florianópolis, Brazil, 2008.

11. Da Silva, F.P. *Oxidação Alílica de Alcenos Catalisada por Nanopartículas de Óxido de Cobalto Suportadas*; Universidade de São Paulo: São Paulo, Brazil, 2011.

12. Di Serio, M.; Turco, R.; Pernice, P.; Aronne, A.; Sannino, F.; Santacesaria, E. Valuation of Nb 2O 5-SiO 2 catalysts in soybean oil epoxidation. *Catal. Today* **2012**, *192*, 112–116. [CrossRef]

13. Nogueira, F.G.E.; Lopes, J.H.; Silva, A.C.; Lago, R.M.; Fabris, J.D.; Oliveira, L.C.A. Catalysts based on clay and iron oxide for oxidation of toluene. *Appl. Clay Sci.* **2011**, *51*, 385–389. [CrossRef]

14. Ingle, R.H.; Vinu, A.; Halligudi, S.B. Alkene epoxidation catalyzed by vanadomolybdophosphoric acids supported on hydrated titania. *Catal. Commun.* **2008**, *9*, 931–938. [CrossRef]

15. Gallo, A.; Tiozzo, C.; Psaro, R.; Carniato, F.; Guidotti, M. Niobium metallocenes deposited onto mesoporous silica via dry impregnation as catalysts for selective epoxidation of alkenes. *J. Catal.* **2013**, *298*, 77–83. [CrossRef]

16. Ziolek, M. Catalytic liquid-phase oxidation in heterogeneous system as green chemistry goal—Advantages and disadvantages of MCM-41 used as catalyst. *Catal. Today* **2004**, *90*, 145–150. [CrossRef]

17. Li, L.; Deng, J.; Yu, R.; Chen, J.; Wang, Z.; Xing, X. Niobium pentoxide hollow nanospheres with enhanced visible light photocatalytic activity. *J. Mater. Chem. A* **2013**, *1*, 11894. [CrossRef]

18. Nakajima, K.; Fukui, T.; Kato, H.; Kitano, M.; Kondo, J.N.; Hayashi, S.; Hara, M. Structure and Acid Catalysis of Mesoporous $Nb_2O_5 \cdot nH_2O$. *Chem. Mater.* **2010**, *22*, 3332–3339. [CrossRef]

19. Zhang, H.; Wang, Y.; Yang, D.; Li, Y.; Liu, H.; Liu, P.; Wood, B.J.; Zhao, H. Directly Hydrothermal Growth of Single Crystal $Nb_3O_7(OH)$ Nanorod Film for High Performance Dye-Sensitized Solar Cells. *Adv. Mater.* **2012**, *24*, 1598–1603. [CrossRef] [PubMed]

20. Ristić, M.; Popović, S.; Musić, S. Sol-gel synthesis and characterization of Nb_2O_5 powders. *Mater. Lett.* **2004**, *58*, 2658–2663. [CrossRef]

21. Morandin, M.; Gavagnin, R.; Pinna, F.; Strukul, G. Oxidation of Cyclohexene with Hydrogen Peroxide Using Zirconia–Silica Mixed Oxides: Control of the Surface Hydrophilicity and Influence on the Activity of the Catalyst and Hydrogen Peroxide Efficiency. *J. Catal.* **2002**, *212*, 193–200. [CrossRef]

22. Zhao, H.; Zhou, C.H.; Wu, L.M.; Lou, J.Y.; Li, N.; Yang, H.M.; Tong, D.S.; Yu, W.H. Catalytic dehydration of glycerol to acrolein over sulfuric acid-activated montmorillonite catalysts. *Appl. Clay Sci.* **2013**, *74*, 154–162. [CrossRef]

23. Chagas, P.; Oliveira, H.S.; Mambrini, R.; Le Hyaric, M.; De Almeida, M.V.; Oliveira, L.C.A. A novel hydrofobic niobium oxyhydroxide as catalyst: Selective cyclohexene oxidation to epoxide. *Appl. Catal. A Gen.* **2013**, *454*, 88–92. [CrossRef]

24. Landmesser, H.; Kosslick, H.; Storek, W.; Fricke, R. Interior surface hydroxyl groups in ordeted mesoporous silicates. *Solid State Ion.* **1997**, *76*, 271–277. [CrossRef]

25. Amgarten, D.R. *Determinação do Volume Específico de Poros de Sílicas Cromatográficas por Dessorção de Líquidos em Excesso*; Universidade Estadual de Campinas: Campinas, Brazil, 2006.

26. Kruk, M.; Jaroniec, M. Gas adsorption characterization of ordered organic-inorganic nanocomposite materials. *Chem. Mater.* **2001**, *13*, 3169–3183. [CrossRef]

27. De Souza, W.F.; Guimarães, I.R.; Guerreiro, M.C.; Oliveira, L.C.A. Catalytic oxidation of sulfur and nitrogen compounds from diesel fuel. *Appl. Catal. A Gen.* **2009**, *360*, 205–209. [CrossRef]

28. Ziolek, M.; Sobczak, I.; Decyk, P.; Sobańska, K.; Pietrzyk, P.; Sojka, Z. Search for reactive intermediates in catalytic oxidation with hydrogen peroxide over amorphous niobium(V) and tantalum(V) oxides. *Appl. Catal. B Environ.* **2015**, *164*, 288–296. [CrossRef]

© 2018 by the authors. Licensee MDPI, Basel, Switzerland. This article is an open access article distributed under the terms and conditions of the Creative Commons Attribution (CC BY) license (http://creativecommons.org/licenses/by/4.0/).

applied
sciences

MDPI

Article

Controlling Chemical Reactions in Confined Environments: Water Dissociation in MOF-74

Erika M. A. Fuentes-Fernandez [1,†], Stephanie Jensen [2,†], Kui Tan [1], Sebastian Zuluaga [2], Hao Wang [3], Jing Li [3], Timo Thonhauser [2,4] and Yves J. Chabal [1,*]

[1] Department of Materials Science and Engineering, University of Texas at Dallas, Richardson, TX 75080, USA; erika.fuentesf@gmail.com (E.M.A.F.-F.); kuitan@utdallas.edu (K.T.)
[2] Department of Physics and Center for Functional Materials, Wake Forest University, Winston-Salem, NC 27109, USA; jensensj@wfu.edu (S.J.); sebastian.zuluaga@vanderbilt.edu (S.Z.); thonhauser@wfu.edu (T.T.)
[3] Department of Chemistry and Chemical Biology, Rutgers University, Piscataway, NJ 08854, USA; whwhuccms@gmail.com (H.W.); jingli@rutgers.edu (J.L.)
[4] Department of Chemistry, Massachusetts Institute of Technology, Cambridge, MA 02139, USA
* Correspondence: chabal@utdallas.edu; Tel.: +1-972-883-5751
† These authors contributed equally to this work.

Received: 5 January 2018; Accepted: 1 February 2018; Published: 12 February 2018

Abstract: The confined porous environment of metal organic frameworks (MOFs) is an attractive system for studying reaction mechanisms. Compared to flat oxide surfaces, MOFs have the key advantage that they exhibit a well-defined structure and present significantly fewer challenges in experimental characterization. As an example of an important reaction, we study here the dissociation of water—which plays a critical role in biology, chemistry, and materials science—in MOFs and show how the knowledge of the structure in this confined environment allows for an unprecedented level of understanding and control. In particular, combining in-situ infrared spectroscopy and first-principles calculations, we show that the water dissociation reaction can be selectively controlled inside Zn-MOF-74 by alcohol, through both chemical and physical interactions. Methanol is observed to speed up water dissociation by 25% to 100%, depending on the alcohol partial pressure. On the other hand, co-adsorption of isopropanol reduces the speed of the water reaction, due mostly to steric interactions. In addition, we also investigate the stability of the product state after the water dissociation has occurred and find that the presence of additional water significantly stabilizes the dissociated state. Our results show that precise control of reactions within nano-porous materials is possible, opening the way for advances in fields ranging from catalysis to electrochemistry and sensors.

Keywords: metal organic framework; reaction mechanism; confined environment

1. Introduction

The assembly of water at the interface of materials has become one of the central topics in biology, chemistry, and materials sciences [1]. It can lead to catalytic reactions at oxide surfaces [2–4] and has also attracted considerable interest due to promising applications in photocatalysis, electrochemistry, and sensors [5–8]. Although understanding the mechanism of water dissociation at surfaces is of paramount interest, its study on flat oxide surfaces is unfortunately highly non-trivial—mostly due to experimental challenges in structural characterization and high pressure required (on the order of 20 atm) for in-situ studies on flat surfaces [1,7,9–12]. In this respect, metal organic frameworks (MOFs) are attractive systems to study reactions since they are well-controlled crystalline environments with well-defined metal oxide centers [3,4,13,14], in which high gas densities can be achieved at relatively low external pressures. MOFs are networks of metal ions linked by organic ligands, forming a well-characterized cross-linked structure. Due to their porous nature, gas molecules can penetrate deeply into the MOF network [15] and experience adsorption forces that allow for significantly longer residence times than at surfaces under similar pressures and temperatures, thus fostering an environment much more conducive to trigger, observe, and study desired reactions.

MOFs are already well studied for their high surface areas and nano-porous structure as they provide an ideal environment for many applications ranging from gas storage and separation to sensors and catalysis [16–20]. Amongst the vast amount of existing MOFs, MOF-74 is of great interest since it contains a high density of coordinately unsaturated metal centers (also called open metal sites) in metal-oxide pyramid clusters, which act as active adsorption sites for many small molecules such as H_2, CO, CO_2, NO, CH_4, and H_2O [21–29]. In particular, Zn-MOF-74 exhibits a strong affinity to water, making it ideal for examining various water reaction mechanisms [30]. We have already studied the reaction of water molecules alone in Zn-MOF-74 [2,3,14,31], finding direct evidence for water dissociation at only 150 °C—this was achieved through the use of D_2O instead of H_2O, observing a clear fingerprint peak at 970 cm^{-1} that corresponds to a O–D bending vibration of OD groups formed upon deuteration of the organic linker [3]. The concerted action between open metal sites and linker phenolate group plays an essential role in breaking up water molecules since water molecule establish strong coordinative bond with metal center via its oxygen and hydrogen bonding interaction with nearby –C–O– moiety from organic linker. With an increase in temperature the water molecule dissociates into D and OD [3]. The OD binds to the open metal sites, while the D atom is transferred to the oxygen of phenolate group. The reaction with water (H_2O) itself follows of course the same dissociation pathway, but spectroscopic signature is impossible to detect with H_2O as hydrogen blue shifts the peak outside the phonon gap of the MOF, where it strongly couples to many other modes. It was also demonstrated that the formation of water networks within the MOF pores was crucial to further lower the corresponding reaction barrier via a proton-exchange mechanism and that physical obstruction of this network with inert molecules (He, Ar) hinders the reaction [14].

In the present work, we explore the role of co-adsorbed alcohol molecules and present a kinetic analysis of the water dissociation reaction inside Zn-MOF-74 in the presence of additional guest molecules. We find that alcohol molecules affect the dissociation rate, either by enhancing or blocking the reaction depending on the type of interaction resulting from such co-adsorption. Specifically, methanol enhances the reaction due to H-bonding interactions, while isopropyl alcohol (IPA) hinders it because steric interactions dominate in that case. The knowledge derived from these combined studies is directly relevant to fields such as catalysis and sensors by providing a means to control water-related reactions, as well as to applications such as gas storage and separation by suppressing water dissociation and therefore extending the MOF lifetime.

2. Methods

2.1. Sample Preparation and In-Situ Infrared Spectroscopy

Zn-MOF-74 powder (1.5 mg) was gently pressed onto a KBr pellet and placed inside a high-pressure cell (Specac Ltd., Orpington, UK; product number P/N 5850c). This high-pressure cell was located in the sample compartment of a Nicolet 6700 FTIR spectrometer (Thermo Scientific Inc., Mountain View, CA, USA) with the sample at the focal point of the beam. All the infrared (IR) spectroscopic data were collected with a liquid N_2-cooled mercury cadmium telluride (MCT-B or MCT-A from Thermo Scientific Inc., Mountain View, CA, USA) detector. The cell was connected to different gas lines for exposure and a vacuum line for evacuation. All spectra were recorded in transmission mode from 400 cm^{-1} (MCT-B) or 650 cm^{-1} (MCT-A) to 4000 cm^{-1} (4 cm^{-1} spectral resolution). Regular water (H_2O) and heavy water (D_2O) were used, with most of the work performed with D_2O to avoid a MOF phonon overlap in the spectral range of the O–H bending vibration [2,3].

2.2. Experimental Measurement Conditions

Once inside the high-pressure cell, the MOF powder was activated by annealing at 180 °C for at least 4 h in vacuum (<50 mTorr) to remove the solvent and ambient humidity from the inside of MOF's pores. The sample was then cooled down to room temperature. In a mixing gas chamber connected to the high-pressure cell (where the sample is located). 8 Torr of alcohol (methanol or isopropyl alcohol) was prepared with 8 Torr of deuterated water (D_2O). The alcohol/D_2O mixture was then introduced into the main chamber and allowed to stabilize for 10 min, after which the temperature was raised to 180 °C. Spectra were recorded as a function of time until stabilization of the 970 cm^{-1} peak was reached (~8 h). For completeness, the results are compared to data previously reported by our group using He and pure D_2O [14]. Measurements with pure MeOH and its deuterated analog (MeOD) are also presented as reference in the Supplementary Information (See Figure S2).

2.3. Computational Details

Ab initio modeling was performed at the density functional theory (DFT) level in Vienna Ab initio simulation package (VASP) [32,33]. To include van der Waals interactions of guest molecules within the MOF, the exchange-correlation functional vdW-DF was used [34–37]. The energy cutoff was set to 600 eV and only the Γ point was considered due to the size of the system. The 54 atom rhombohedral primitive cell of Zn-MOF-74 was first relaxed until the forces on the atoms were below 1 meV/Å. Binding energies were then calculated for MeOH, H_2O, OH, and H (and deuterated equivalents), with the hydrogen or deuterium at the metal center and on the oxygen atom of the MOF organic linker. After the guest molecules were placed into the MOF, each structure was again relaxed until the forces on all atoms were below 1 meV/Å. Reaction barriers were calculated with a standard transition-state search algorithm, i.e., the nudged-elastic band method [38,39], as implemented in VASP.

3. Results and Discussion

3.1. Experimental Quantification of the Water Dissociation Reaction Rate

To determine the impact of guest molecules on the water dissociation rate through chemical interactions, we studied this reaction in the presence of alcohol molecules inside the MOF as a function of temperature and time. After MOF activation, the alcohol/D_2O mixture was introduced into the cell at room temperature and IR absorption spectra were subsequently recorded with the sample at 180 °C, monitoring the above-mentioned O–D bending vibration at ~970 cm^{-1}, which is the product of the D_2O dissociation [3]. The IR spectra were recorded until the intensity of the 970 cm^{-1} peak stabilized. In this manner, the kinetics of methanol/D_2O and isopropanol/D_2O mixtures could be compared for different alcohol pressures with previously reported data with either pure D_2O or inert gas/D_2O mixtures [14], using the integrated area of the 970 cm^{-1} peak as a quantification of D_2O dissociation. Results with pure methanol and pure deuterated methanol are also presented for comparison in the supplementary information (see Figure S2).

Figure 1a shows the representative water dissociation fingerprint at 970 cm^{-1} after 8 h of reaction at 180 °C for the different mixtures analyzed inside the MOF, which is sufficient time for the reaction to stabilize in all environments. All spectra are normalized to the quantity (weight) of the MOF powder. The feature at 1003 cm^{-1}, only observed for the mixture with 8 Torr MeOH, will be discussed shortly. Figure 1b shows the kinetic evolution of the tested reactions. It is clear that water dissociation rates strongly depend on the environment: the $D_2O \rightarrow 2D + O$ reaction is faster in the presence of methanol and slower in the presence of isopropanol or inert gases such as He. Experimentally, we find that the water reaction rates inside MOF-74 are fastest to slowest as follows: 8 Torr MeOH > 4 Torr MeOH > Pure D_2O > 8 Torr IPA > 950 Torr He. Table 1 reports the percent of dissociated water molecules, as compared with pure water (100%), as function of guest molecules. These data can be used to examine the dependence on the alcohol pressure, for instance in the case of methanol. The dependence is clearly not linear, but can be well fitted with a quadratic fit, $y = 25.5x^2 - 149.5x + 219$, where y is the percent dissociation and x the pressure, pointing to the importance of the local environment. Indeed, water dissociation is faster when there are more MeOH molecules inside the MOF pore, i.e., when a network can be formed.

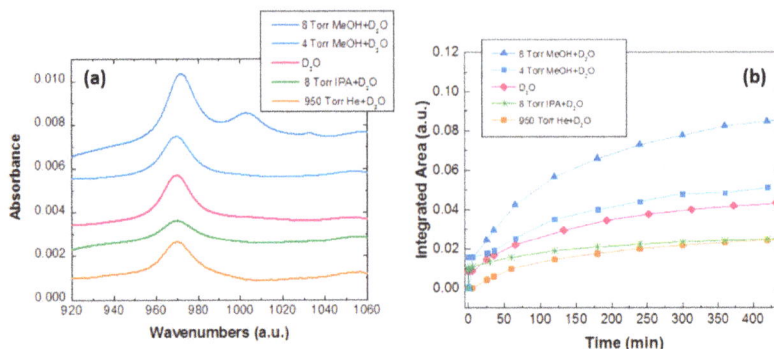

Figure 1. (**a**) Infrared spectra of Zn-MOF-74 impregnated with the different alcohol mixtures (methanol = Me and Isopropyl alcohol = IPA) for 8 h at 180 °C, compared to a He/D$_2$O mix and pure D$_2$O previously reported [14]. (**b**) Kinetic plots of the dissociation reaction over time (integrated area of 970 cm^{-1} band). All spectra are normalized to metal organic frameworks (MOF) quantity.

Table 1. Percentage of deuterated water dissociated after 8 h of reaction for D$_2$O + guest molecules, based on data shown in Figure 1b, compared to pure D$_2$O (taken as the reference, i.e., 100%).

D$_2$O + Guest Molecules	Percentage of D$_2$O Dissociated after 8 h (%)
8 Torr D$_2$O	100
4 Torr MeOH + 8 Torr D$_2$O	122
8Torr MeOH + 8 Torr D$_2$O	195
8 Torr IPA + 8 Torr D$_2$O	61
950 Torr He + 8 Torr D$_2$O	61

Methanol is interesting, namely because of its small size that minimizes steric hindrance within the MOF pore. Different pressures of methanol were tested while keeping the D$_2$O pressure at 8 Torr for all experiments. The infrared spectra measured after introduction of a MeOH/D$_2$O mixture as a function of temperature and time are presented in Figure 2. After introducing a methanol/D$_2$O mixture into Zn-MOF-74 at room temperature, the C–O stretch mode of methanol gas phase was observed at 1033 cm^{-1} (see Figure S3). As the temperature increases, this band red-shifts to 1010 cm^{-1} at 170 °C and to 1003 cm^{-1} at 180 °C, which suggests a strong interaction of methanol with the MOF metal center [40].

Water dissociation only proceeds at 180 °C, as evidenced by the appearance of the 970 cm^{-1} peak. At that point, the peak at 1003 cm^{-1} weakens, indicating that C–O dissociation occurs. The synergistic dissociation of methanol and D$_2$O results in a faster D$_2$O dissociation reaction. In fact, the increased dissociation with MeOH pressure described earlier suggests that the presence of methanol fosters the reaction, mediated by the metal site, almost doubling the reaction yield in 8 Torr MeOH. This observation can be rationalized with first-principles calculations as detailed next.

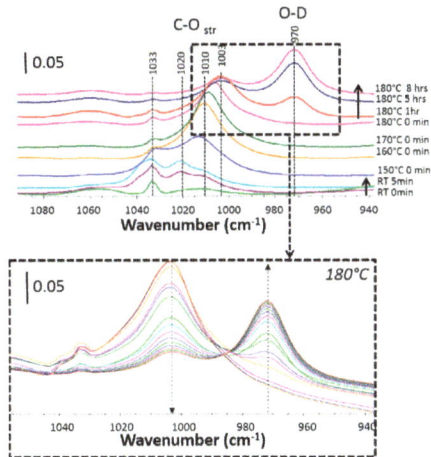

Figure 2. (**Top**) Absorbance IR absorption spectra of Zn-MOF-74 loaded with a mixture of 8 Torr methanol and 8 Torr D_2O as a function of temperature and time. (**Bottom**) Detailed reaction evolution at 180 °C from 0 to 8 h (dashed arrows indicate the time evolution).

3.2. Theoretical Analysis of the Water Dissociation Reaction Rate

Calculations summarized in Table 2 show that, when both water and methanol are present in the system, methanol associates to the metal center and water binds to the organic linker. Water alone can associate with both the metal linker and organic linker almost equally, but the MeOH has a stronger binding (association) to the metal center than water. In fact, MeOH does not interact with the oxygen atom of the linker, in contrast to water. Consequently, the OH group of MeOH binds to the metal center upon cooperative adsorption of MeOH and D_2O, while the D atom from water binds to the oxygen atom of the linker since the metal site is occupied, and the OD part of D_2O binds to the CH_3 group of MeOH, as illustrated in Figure 3a. The dissociation reaction then proceeds as follows: First, there is a lengthening and thus weakening of the C–O bond in MeOH, which is in agreement with the observed redshift of $\nu(C–O)$ from 1033 cm^{-1} to 1003 cm^{-1} and facilitates subsequent OH exchange. Indeed, at the transition state, the OD group from the neighboring D_2O molecule approaches the methyl group, while the other D atom bonds to the oxygen of the ligand. As OH (from methanol) bonds to the Zn center very strongly (see Table 2) and D (from D_2O) binds strongly to O (see Table 2), the OD group is transferred to the methyl group to complete the hydroxyl exchange on the methyl group. While such an exchange is not common, the strong energetic gain drives this reaction. As MeOD is formed, a single D remains at the oxygen of the linker, characterized by the well-studied 970 cm^{-1} IR fingerprint [3]. Note that in this cooperative reaction the energy barrier for dissociation is also reduced by a clustering effect [14]. The products of the dissociation reaction (OH on the Zn^{2+} site or D on oxygen site) bind much more strongly than the initial reactants (H_2O/D_2O and MeOH) that are only bound through weaker van der Waals interactions, as shown in Table 2.

Table 2. Calculated binding energies (in eV) of reactants and products to both the Zn metal center and the organic linker prior to and after water dissociation for a methanol/water mixture. DNB indicates that the functional did not bind.

Binding Energies of Reactant and Products Inside Zn-MOF-74		
Functional	Zn Metal Center E_b (eV)	Organic Linker E_b (eV)
MeOH	−0.780	DNB
H_2O	−0.612	−0.610
OH	−2.442	DNB
H	−2.002	−1.932

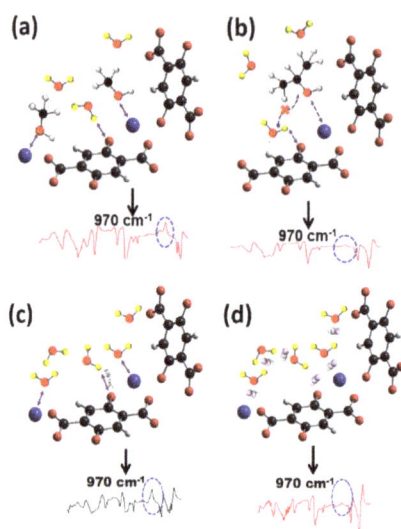

Figure 3. Schematic of Zn-MOF-74 interacting with (**a**) methanol/D_2O assisting the reaction by the formation of clusters in a similar fashion as pure D_2O; (**b**) isopropanol/D_2O hindering the interaction of D_2O with the secondary carbon in the IPA molecule due to the steric barrier; (**c**) pure deuterated water (D_2O) and cluster formation; and (**d**) He/D_2O, with He obstructing water cluster formation. White, yellow, red black, blue and purple spheres represent H, D, O, C, Zn, and He atoms. Each schematic has a normalized Zn-MOF-74 IR spectra after 8 h of reaction at 180 °C.

The mechanism by which MeOH can thus speed up the reaction is as follows: When water dissociates without any other guest molecules present, it is the water molecule bound to the metal center that dissociates, as shown in Figure 3c [3,14,30,31]. In the presence of MeOH, this pathway is blocked, as MeOH takes the place on the metal site, which forces the water bound to the linker, as shown in Figure 3a. The weakening of C–O bond in MeOH, evidenced by the frequency shift, is even more pronounce when an OD group (from D_2O) is in close proximity, which makes MeOH an effective precursor for D_2O dissociation, leading to D bonding to the oxygen linker, with reformation of a methanol (MeOD) molecule. To further support the experimental observation of a faster reaction, the vibrational modes of the D_2O/MeOH system were also calculated in conjunction with Zn-MOF-74 and characterized using J-ICE [41]. After applying a systematic scaling of 7% to all computed frequencies, a peak at 1030 cm^{-1} was found for the C–O stretching mode of gas-phase MeOH; that mode was also calculated to red-shift to 1005 cm^{-1} upon adsorption in the MOF at the Zn metal centers and interaction with D_2O. This calculated 25 cm^{-1} red shift is in excellent agreement with the measured 30 cm^{-1} redshift, i.e., from 1033 cm^{-1} to 1003 cm^{-1}, mentioned above. As pointed out, this red-shift is indicative of the C–O bond weakening, which is a requirement for hydroxyl exchange MeOH → MeOD. The disappearance of the 1003 cm^{-1} mode in Figure 2b thus indicates that more and more MeOH gets converted into MeOD (1060 cm^{-1}) via the mechanism discussed above.

Isopropyl alcohol, on the other hand, slows down the dissociation reaction over time (Figure 1b), similarly to what is observed for the He/D$_2$O mixture (Figure 3d), suggesting that steric interactions are important in the case of IPA. We previously reported that inert gases such as He and Ar block the intermolecular interactions between water molecules by disrupting the formation of water clusters (networks), thus effectively increasing the reaction barrier [14]. Similarly to noble gases, IPA prevents water cluster formation, even at much lower pressure than noble gases, thus slowing down the dissociation rate. In contrast to methanol that fosters the reaction, the size of IPA prevents reaction more than the propensity of its OH group to react with the metal site, as discussed below. Note that the chemical and crystalline structures of MOF-74 remain after D$_2$O/alcohol exposure (see Figure S7).

In summary, the experimental results supported by computational analysis suggest the reactions illustrated in Figure 3 (with reaction pathways simulated for 1 vs. 4 water molecules reacting with 1 MeOH molecule is presented in Supplementary Materials in Section S4). Methanol by itself has a higher steric hindrance than water, but not high enough to block the reaction given the overall pore size in MOF-74. Because the reaction is experimentally observed to take place at high temperature, dissociation of methanol is favored; as a consequence of MeOH's stronger interaction with the metal center, the reaction achieves a more stable final state (Table 2). After interacting with the metal center, MeOH and D$_2$O molecules form an organized structure (cluster), which helps to reduce the dissociation reaction barrier of D$_2$O at the organic linker—and the reaction proceeds as outlined above. Steric obstruction of water clustering due to a bulkier alcohol, such as isopropanol, is presented in Figure 3b. The −CH$_3$ groups surrounding the secondary carbon block water molecules from clustering and reduce the dissociation rate by 40% in comparison to pure D$_2$O as presented in Table 1. Note that a lower pressure (8 Torr) of IPA is needed than He gas (950 Torr) to reduce the water dissociation rate to the same level, suggesting co-adsorbing IPA is an effective way to suppress water dissociation and therefore to increase the MOF stability for gas uptake and other applications. Further details are presented in the supplementary information in Section S5.

Comparing the two pathways that obstruct the reaction, a schematic of the reaction mechanism of IPA/water is presented in Figure 3b. The IPA blocks the reaction by its interaction with the metal center, requiring a lower pressure to occupy all metal centers in the unit cell [42]. On the contrary He/water reaction mechanism presented in Figure 3d illustrates how adding guest molecules physically prevent the formation of organized water clusters, hence reducing water dissociation reaction at higher pressures. Thus, the water dissociation reaction can be simply controlled by assisting or hampering cluster formation by a guest molecule such as methanol or isopropanol, therefore systematically tuning the effective reaction barriers [14].

For both pure water and alcohol/water mixtures, a central aspect of MOF-74 is the ability to adsorb several molecules within each pore at relatively low external pressures (~few Torr), thanks to van der Waals attractive forces. This is only possible in the confined space of a nanoporous material. In the confined environment of MOF-74, several water and alcohol molecules can coexist within each pore, forming nano-structures, which generally lowers the dissociation barriers and stabilizes the dissociated state. Consequently, the formation of water or alcohol/water networks occurs at convenient pressures, and these networks can remain stable even after the cell is pumped out. Such networks cannot be realized on flat oxide surfaces because the external pressure would need to be prohibitively high and because entropy would prevent structured networks to develop for lack of a supporting nanostructured framework around [11,12].

3.3. Stabilization Effect of Additional Water on the Water Dissociation Reaction

A final point needs to be addressed, i.e., the stability of the dissociation products, because the reverse reaction barriers tend to be low. A quantitative analysis is difficult because reaction barriers are difficult to evaluate in systems where there may be several reaction pathways. Nonetheless, we have attempted such analysis by calculating the water dissociation reaction barrier using a standard transition-state search algorithm [38,39], in the presence of a varying amount of additional water molecules. In particular, we investigate the reaction mechanism involving MeOH assistance as discussed above, and compare that with the dissociation of pure H_2O found previously [14]. Results are presented in Table 3 and Figure 4.

Table 3. Reaction barriers for the MeOH assisted and pure H_2O dissociation in the presence of varying amounts of additional water molecules.

Water Systems	No. H_2O	Initial State (eV)	Transition State (eV)	Final State (eV)	Forward Barrier (eV)	Backward Barrier (eV)
MeOH + H_2O	1	0	1.607	1.386	1.607	0.220
	4	0	0.932	0.598	0.932	0.333
H_2O	1	0	1.085	1.054	1.085	0.031
	2	0	0.964	0.832	0.964	0.132
	3	0	0.990	0.571	0.990	0.418
	4	0	0.688	0.530	0.688	0.158
	5	0	0.690	0.580	0.690	0.110

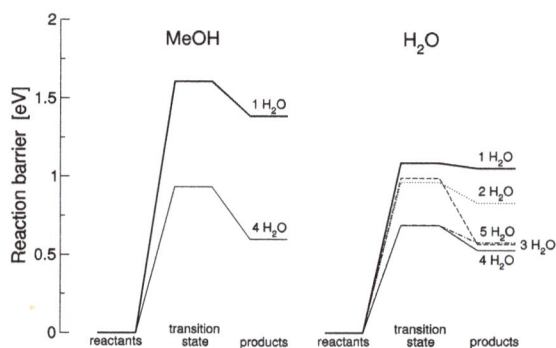

Figure 4. Reaction barriers for the MeOH assisted and pure H_2O dissociation in the presence of varying amounts of additional water molecules.

In both cases—the MeOH assisted and pure H_2O dissociation—it is obvious that increasing the number of water molecules decreases the forward reaction barrier for the individual reactions. This behavior has been analyzed in detail for the H_2O case, where it was found that small water clusters form and are, in turn, responsible for decreasing the barrier [14]; we propose a similar mechanism in the presence of MeOH. However, more importantly, increasing the number of water molecules also increases the backward barrier, i.e., the barrier for the reverse reaction. In fact, in some cases (such as the pure water dissociation with only one H_2O) the reverse barrier is so low that the product state cannot be considered stable at all and, even at low temperature, will recombine back into water molecules. It is thus the presence of additional water molecules that stabilizes the product state and makes the reaction observable. Note that experimentally, it is very difficult to determine the exact amount of water molecules inside the MOF pore and nearby the reaction site, but it is estimated based on isotherm measurements corrected to the higher temperature (180 °C), that 8 Torr of water result in at least 6 to 8 water molecules in the MOF unit cell (calculations based on the ideal gas law) [42].

In the previous section, comparison between pure water and water/alcohol mixtures was based on total energy calculations, providing good support for experimental observations. Calculating reaction barriers through transition-state searches becomes increasingly more complex in the presence of more water molecules (due to the vastly growing space of possible arrangements and reaction pathways) so that such calculations only provide upper limits for reaction barriers. Due to their complexity and computer time needed, we only investigated a limited set. For instance, barriers for a MeOH/water mixture are only calculated for one and four water molecules. Based on calculations presented in Figure 4, it is suggested that water dissociation does not proceed with only one water molecule. It appears that, in the case of MeOH the forward barrier is too high to account for the experimentally observed reaction at 180 °C; also, in the case of a single H_2O molecule, the backward barrier is so small that it would not be possible to observe the reaction before it reverses. In the H_2O case, the largest backwards barrier (0.418 eV) is found for three water molecules, but that scenario does not have the smallest forward barrier. While details are difficult to model and quantify, the results suggest that the presence of a small network of water molecules lowers the forward reaction, and the presence of water molecules remaining after the reaction stabilizes the products. In reality, there must be a dynamic equilibrium that gets established for given pressures. Therefore, this section only provides a suggestion for the mechanisms involved yet cannot unambiguously quantify them.

4. Conclusions

Combining in-situ infrared spectroscopic studies and first-principles DFT calculations, we have shown that the dissociation reaction of water molecules within MOF-74 can be controlled by co-adsorption of alcohols, just as it is hindered by inert gases. Bulky alcohol molecules such as IPA prevent water clustering, thus reducing water dissociation rates more effectively than inert gases at comparable external pressures. On the other hand, small alcohol molecules with a low boiling point, such as methanol, effectively accelerate water dissociation, without sterically hindering the process. In the confined environment of MOF-74, several water molecules can coexist within each pore, forming nano-structures, which generally lowers the dissociation barriers and stabilizes the dissociated state at low pressures. These results offer a potential approach to controlling water dissociation inside MOF-74, which is important for gas storage and separation applications. Since water is present in most biological, catalytic, and daily life applications, the understanding derived here provides new insight into the possibility of controlling water dissociation rates, which in turn affords a novel means for tuning chemical reactions in general.

Supplementary Materials: The following are available online at www.mdpi.com/2076-3417/8/2/270/s1. Sample preparation and activation methods; Powder X-ray diffraction pattern of Zn-MOF-74 sample; Isotherm measurement of N_2 adsorption into Zn-MOF-74; IR spectra of pure methanol (MeOH) and deuterated methanol (MeOD) interaction with Zn-MOF-74; Water dissociation reaction pathway in the presence of one and four water molecules; IR spectra of isopropanol (IPA) and D_2O interaction with Zn-MOF-74. This material is available free of charge at http://pubs.acs.org.

Acknowledgments: This work was entirely supported by the Department of Energy Grant No. DE-FG02-08ER46491. T.T. also acknowledges generous support from the Simons Foundation through Grant No. 391888, which endowed his sabbatical leave at MIT.

Author Contributions: H.W. and J.L. synthesized and characterized Zn-MOF-74 used in this work; E.M.A.F.-F., K.T. and Y.J.C. conceived, designed and perform the experimental part of the paper; S.J., S.Z. and T.T. performed the theoretical analysis. All authors contributed to the writing, and analysis of this paper.

Conflicts of Interest: The authors declare no conflict of interest.

References

1. Henderson, M.A. The interaction of water with solid surfaces: Fundamental aspects revisited. *Surf. Sci. Rep.* **2002**, *46*, 1–308. [CrossRef]
2. Tan, K.; Nijem, N.; Gao, Y.; Zuluaga, S.; Li, J.; Thonhauser, T.; Chabal, Y.J. Water interactions in metal organic frameworks. *CrystEngComm* **2015**, *17*, 247–260. [CrossRef]
3. Tan, K.; Zuluaga, S.; Gong, Q.; Canepa, P.; Wang, H.; Li, J.; Chabal, Y.J.; Thonhauser, T. Water Reaction Mechanism in Metal Organic Frameworks with Coordinatively Unsaturated Metal Ions: MOF-74. *Chem. Mater.* **2014**, *26*, 6886–6895. [CrossRef]
4. Ward, M.D. Confined systems: The bright side of MOFs. *Nat. Chem.* **2010**, *2*, 610–611. [CrossRef] [PubMed]
5. Hass, K.C.; Schneider, W.F.; Curioni, A.; Andreoni, W. The Chemistry of Water on Alumina Surfaces: Reaction Dynamics from First Principles. *Science* **1998**, *282*, 265–268. [CrossRef] [PubMed]

6. Brown, G.E., Jr. How Minerals React with Water. *Science* **2001**, *294*, 67–69. [CrossRef] [PubMed]
7. Bikondoa, O.; Pang, C.L.; Ithnin, R.; Muryn, C.A.; Onishi, H.; Thornton, G. Imaging water dissociation on TiO2(110). *Nat. Mater.* **2006**, *5*, 189–192. [CrossRef]
8. Song, Z.; Fan, J.; Xu, H. Strain-induced water dissociation on supported ultrathin oxide films. *Sci. Rep.* **2016**, *6*, 22853–22858. [CrossRef] [PubMed]
9. Giordano, L.; Goniakowski, J.; Suzanne, J. Partial Dissociation of Water Molecules in the (3 × 2) Water Monolayer Deposited on the MgO (100) Surface. *Phys. Rev. Lett.* **1998**, *81*, 1271–1273. [CrossRef]
10. Odelius, M. Mixed Molecular and Dissociative Water Adsorption on MgO[100]. *Phys. Rev. Lett.* **1999**, *82*, 3919–3922. [CrossRef]
11. Carrasco, J.; Hodgson, A.; Michaelides, A. A molecular perspective of water at metal interfaces. *Nat. Mater.* **2012**, *11*, 667–674. [CrossRef] [PubMed]
12. Rodriguez, J.; Goodman, D.W. High-pressure catalytic reactions over single-crystal metal surfaces. *Surf. Sci. Rep.* **1991**, *14*, 1–107. [CrossRef]
13. Tan, K.; Canepa, P.; Gong, Q.; Liu, J.; Johnson, D.H.; Dyevoich, A.; Thallapally, P.K.; Thonhauser, T.; Li, J.; Chabal, Y.J. Mechanism of Preferential Adsorption of SO2 into Two Microporous Paddle Wheel Frameworks M(bdc)(ted)0.5. *Chem. Mater.* **2013**, *25*, 4653–4662. [CrossRef]
14. Zuluaga, S.; Fuentes-Fernandez, E.M.; Tan, K.; Li, J.; Chabal, Y.J.; Thonhauser, T. Cluster assisted water dissociation mechanism in MOF-74 and controlling it using helium. *J. Mater. Chem. A* **2016**, *4*, 11524–11530. [CrossRef]
15. Canepa, P.; Nijem, N.; Chabal, Y.J.; Thonhauser, T. Diffusion of Small Molecules in Metal Organic Framework Materials. *Phys. Rev. Lett.* **2013**, *110*, 026102. [CrossRef] [PubMed]
16. Suh, M.P.; Park, H.J.; Prasad, T.K.; Lim, D.-W. Hydrogen Storage in Metal–Organic Frameworks. *Chem. Rev.* **2011**, *112*, 782–835. [CrossRef] [PubMed]
17. Sumida, K.; Rogow, D.L.; Mason, J.A.; McDonald, T.M.; Bloch, E.D.; Herm, Z.R.; Bae, T.-H.; Long, J.R. Carbon Dioxide Capture in Metal–Organic Frameworks. *Chem. Rev.* **2011**, *112*, 724–781. [CrossRef] [PubMed]
18. Li, J.-R.; Sculley, J.; Zhou, H.-C. Metal–Organic Frameworks for Separations. *Chem. Rev.* **2012**, *112*, 869–932. [CrossRef] [PubMed]
19. Kreno, L.E.; Leong, K.; Farha, O.K.; Allendorf, M.; van Duyne, R.P.; Hupp, J.T. Metal–Organic Framework Materials as Chemical Sensors. *Chem. Rev.* **2011**, *112*, 1105–1125. [CrossRef] [PubMed]
20. Wu, H.H.; Gong, Q.H.; Olson, D.H.; Li, J. Commensurate Adsorption of Hydrocarbons and Alcohols in Microporous Metal Organic Frameworks. *Chem. Rev.* **2012**, *112*, 836–868. [CrossRef] [PubMed]
21. Kizzie, A.C.; Wong-Foy, A.G.; Matzger, A.J. Effect of Humidity on the Performance of Microporous Coordination Polymers as Adsorbents for CO2 Capture. *Langmuir* **2011**, *27*, 6368–6373. [CrossRef] [PubMed]
22. Dietzel, P.D.C.; Johnsen, R.E.; Fjellvag, H.; Bordiga, S.; Groppo, E.; Chavan, S.; Blom, R. Adsorption properties and structure of CO2 adsorbed on open coordination sites of metal-organic framework Ni2(dhtp) from gas adsorption, IR spectroscopy and X-ray diffraction. *Chem. Commun.* **2008**, *0*, 5125–5127. [CrossRef] [PubMed]
23. Wu, H.; Zhou, W.; Yildirim, T. High-Capacity Methane Storage in Metal Organic Frameworks M2(dhtp): The Important Role of Open Metal Sites. *J. Am. Chem. Soc.* **2009**, *131*, 4995–5000. [CrossRef] [PubMed]
24. Liu, Y.; Kabbour, H.; Brown, C.M.; Neumann, D.A.; Ahn, C.C. Increasing the Density of Adsorbed Hydrogen with Coordinatively Unsaturated Metal Centers in Metal Organic Frameworks. *Langmuir* **2008**, *24*, 4772–4777. [CrossRef] [PubMed]
25. Bloch, E.D.; Hudson, M.R.; Mason, J.A.; Chavan, S.; Crocellà, V.; Howe, J.D.; Lee, K.; Dzubak, A.L.; Queen, W.L.; Zadrozny, J.M. Reversible CO Binding Enables Tunable CO/H2 and CO/N2 Separations in Metal–Organic Frameworks with Exposed Divalent Metal Cations. *J. Am. Chem. Soc.* **2014**, *136*, 10752–10761. [CrossRef] [PubMed]
26. Bonino, F.; Chavan, S.; Vitillo, J.G.; Groppo, E.; Agostini, G.; Lamberti, C.; Dietzel, P.D.; Prestipino, C.; Bordiga, S. Local Structure of CPO-27-Ni Metallorganic Framework upon Dehydration and Coordination of NO. *Chem. Mater.* **2008**, *20*, 4957–4968. [CrossRef]
27. Canepa, P.; Arter, C.A.; Conwill, E.M.; Johnson, D.H.; Shoemaker, B.A.; Soliman, K.Z.; Thonhauser, T. High-throughput screening of small-molecule adsorption in MOF. *J. Mater. Chem. A* **2013**, *1*, 13597–13604. [CrossRef]

28. Tan, K.; Zuluaga, S.; Fuentes, E.; Mattson, E.C.; Veyan, J.-F.; Wang, H.; Li, J.; Thonhauser, T.; Chabal, Y.J. Trapping gases in metal organic frameworks with a selective surface molecular barrier layer. *Nat. Commun.* **2016**, *7*, 13871–13879. [CrossRef] [PubMed]

29. Tan, K.; Zuluaga, S.; Gong, Q.; Gao, Y.; Nijem, N.; Li, J.; Thonhauser, T.; Chabal, Y.J. Competitive Coadsorption of CO_2 with H_2O, NH_3, SO_2, NO, NO_2, N_2, O_2, and CH_4 in M-MOF-74 (M = Mg, Co, Ni): The Role of Hydrogen Bonding. *Chem. Mater.* **2015**, *27*, 2203–2217. [CrossRef]

30. Zuluaga, S.; Fuentes-Fernandez, E.M.A.; Tan, K.; Arter, C.A.; Li, J.; Chabal, Y.J.; Thonhauser, T. Chemistry in confined spaces: Reactivity of the Zn-MOF-74 channels. *J. Mater. Chem. A* **2016**, *4*, 13176–13182. [CrossRef]

31. Zuluaga, S.; Fuentes-Fernandez, E.M.A.; Tan, K.; Xu, F.; Li, J.; Chabal, Y.J.; Thonhauser, T. Understanding and controlling water stability of MOF-74. *J. Mater. Chem. A* **2016**, *4*, 5176–5183. [CrossRef]

32. Kresse, G.; Furthmüller, J. Efficient iterative schemes for ab initio total-energy calculations using a plane-wave basis set. *Phys. Rev. B* **1996**, *54*, 1169–1186. [CrossRef]

33. Kresse, G.; Joubert, D. From ultrasoft pseudopotentials to the projector augmented-wave method. *Phys. Rev. B* **1999**, *59*, 1758–1775. [CrossRef]

34. Berland, K.; Cooper, V.R.; Lee, K.; Schröder, E.; Thonhauser, T.; Hyldgaard, P.; Lundqvist, B.I. van der Waals forces in density functional theory: A review of the vdW-DF method. *Rep. Prog. Phys.* **2015**, *78*, 066501. [CrossRef] [PubMed]

35. Langreth, D.; Lundqvist, B.I.; Chakarova-Käck, S.D.; Cooper, V.; Dion, M.; Hyldgaard, P.; Kelkkanen, A.; Kleis, J.; Kong, L.; Li, S. A density functional for sparse matter. *J. Phys. Condens. Matter* **2009**, *21*, 084203. [CrossRef] [PubMed]

36. Thonhauser, T.; Cooper, V.R.; Li, S.; Puzder, A.; Hyldgaard, P.; Langreth, D.C. Van der Waals density functional: Self-consistent potential and the nature of the van der Waals bond. *Phys. Rev. B* **2007**, *76*, 125112. [CrossRef]

37. Thonhauser, T.; Zuluaga, S.; Arter, C.; Berland, K.; Schröder, E.; Hyldgaard, P. Spin Signature of Nonlocal Correlation Binding in Metal-Organic Frameworks. *Phys. Rev. B* **2015**, *115*, 136402. [CrossRef] [PubMed]

38. Henkelman, G.; Uberuaga, B.P.; Jonsson, H. A climbing image nudged elastic band method for finding saddle points and minimum energy paths. *J. Chem. Phys.* **2000**, *113*, 9901–9904. [CrossRef]

39. Henkelman, G.; Jónsson, H. Improved tangent estimate in the nudged elastic band method for finding minimum energy paths and saddle points. *J. Chem. Phys.* **2000**, *113*, 9978–9985. [CrossRef]

40. Falk, M.; Whalley, E. Infrared Spectra of Methanol and Deuterated Methanols in Gas, Liquid, and Solid Phases. *J. Chem. Phys.* **1961**, *34*, 1554–1568. [CrossRef]

41. Canepa, P.; Hanson, R.M.; Ugliengo, P.; Alfredsson, M. J-ICE: A new Jmol interface for handling and visualizing crystallographic and electronic properties. *J. Appl. Crystallogr.* **2011**, *44*, 225–229. [CrossRef]

42. Li, Y.; Wang, X.; Xu, D.; Chung, J.D.; Kaviany, M.; Huang, B. H_2O Adsorption/Desorption in MOF-74: Ab Initio Molecular Dynamics and Experiments. *J. Phys. Chem. C* **2015**, *119*, 13021–13031. [CrossRef]

© 2018 by the authors. Licensee MDPI, Basel, Switzerland. This article is an open access article distributed under the terms and conditions of the Creative Commons Attribution (CC BY) license (http://creativecommons.org/licenses/by/4.0/).

applied
sciences

MDPI

Article

The Influence of Quantum Confinement on Third-Order Nonlinearities in Porous Silicon Thin Films

Rihan Wu, Jack Collins, Leigh T. Canham and Andrey Kaplan *

School of Physics and Astronomy, University of Birmingham, Birmingham B15 2TT, UK;
rxw593@bham.ac.uk (R.W.); jdc357@student.bham.ac.uk (J.C.); l.t.canham@bham.ac.uk (L.T.C.)
* Correspondence: a.kaplan.1@bham.ac.uk

Received: 14 September 2018; Accepted: 28 September 2018; Published: 3 October 2018

check for
updates

Abstract: We present an experimental investigation into the third-order nonlinearity of conventional crystalline (c-Si) and porous (p-Si) silicon with Z-scan technique at 800-nm and 2.4-µm wavelengths. The Gaussian decomposition method is applied to extract the nonlinear refractive index, n_2, and the two-photon absorption (TPA) coefficient, β, from the experimental results. The nonlinear refractive index obtained for c-Si is $7 \pm 2 \times 10^{-6}$ cm^2/GW and for p-Si is $-9 \pm 3 \times 10^{-5}$ cm^2/GW. The TPA coefficient was found to be 2.9 ± 0.9 cm/GW and 1.0 ± 0.3 cm/GW for c-Si and p-Si, respectively. We show an enhancement of the nonlinear refraction and a suppression of TPA in p-Si in comparison to c-Si, and the enhancement gets stronger as the wavelength increases.

Keywords: third-order nonlinearity; self-focusing; TPA; porous silicon; Z-scan

1. Introduction

Porous silicon (p-Si) has attracted great interest in recent years due to its unique opto-electronic properties. The nanoscale sponge-like structure of p-Si, whereas porosity significantly enlarges its surface-to-volume ratio, exhibits the quantum confinement effect [1], resulting in faster carrier recombination [2], accompanied, at certain conditions, by photo-luminescence [3] and optical nonlinearities [4]. These effects are usually greatly enhanced in comparison to the conventional crystalline silicon (c-Si). These new capabilities extend the implementation of the traditional silicon material in photonics and optoelectronics to a wider range of potential applications of all-optical switching [5,6], optical sensing [7,8], energy conservation [9] and photonics devices [10]. However, possible applications encountering the enhanced nonlinearities of p-Si can be either an issue or advantage. Unwanted non-linear effects may result in sample damage through the undesirable self-focusing and two-photon absorption effects. However, the same effects may also bring advantages such as the use of p-Si as a material for 3D-laser micro-structuring [11] and an all-optical switching system [12]. Yet, the information on p-Si nonlinearity remains scarce, and our work attempts to form a clearer understanding of the topic through investigating the self-focusing and two-photon absorption (TPA) processes. For the investigation, we employed the Z-scan technique using 800-nm and 2.4-µm femtosecond laser beams.

The Z-scan is the main method used in this work. It was first proposed by Sheikh-Bahae et al. [13] in the 1990s for identifying the third-order nonlinearities of materials by translating a test sample through the focus of a Gaussian beam and recording the transmittance at each sample position. This method is highly sensitive and gives distinguishable features for the measurements of the near (closed aperture) and far (open aperture) field transmittance, corresponding to the nonlinear refraction and TPA processes, respectively. For nonlinearly refractive materials, a higher intensity in the centre of the Gaussian beam, in comparison to its surroundings, results in a variation of the refractive index along the beam cross-section. Therefore, samples behave as a focusing or defocusing lens depending on the sign of the nonlinear refractive index, n_2. This effect can be observed by recording only the centre intensity of the transmitted light, which is normally accomplished by placing an aperture before a detector. In contrast, the total transmitted intensity (without the aperture) carries the information about nonlinear absorption, as it records a change of the total beam intensity, which is commonly governed by the TPA.

The main purpose of this work is to compare nonlinear properties of *p*-Si to *c*-Si and to examine possible effects of the morphology and confinement. We employed the Z-scan method for both *p*-Si and *c*-Si using 800-nm incident light, which has photon energy greater than the band-gap of both materials. The majority of researchers focus on this energy regime because the nonlinear response benefits from strong single-photon resonance enhancement [14]. However, to have a better understanding of the nonlinearities arising from the anharmonic motion of bound electrons, we further extended the experiments to 2.4 μm, where the incident light energy is below the band-gap. We measured the normalised transmittance, $T(z)/T_0$, with a closed and open aperture for *p*-Si and *c*-Si free-standing thin films with a 15- and 9-μm thickness, respectively. Here, $T(z)$ is the measured transmittance at a sample position *z*, relative to the location of the beam focus and T_0 is the linear transmittance. The acquired experimental data were used to retrieve the nonlinear refractive index, n_2, and TPA coefficient, β. For the data analysis, we applied the Gaussian decomposition method [15]. The results suggest a suppression of the TPA, but an enhancement of the nonlinear refraction in *p*-Si. In addition, the closed aperture Z-scan measurements of *p*-Si showed an opposite sign of n_2 in comparison with *c*-Si.

2. Results

For the sample being translated through the focus of a beam with a Gaussian profile and considering only the third order nonlinearity, the total time-integrated transmission, T_{open}, for the open aperture can be expressed as [16–18]:

$$T_{open}(z) = 1 - \frac{1}{2\sqrt{2}} \frac{\beta I_0 L_{eff}}{1 + (\frac{z}{z_0})^2} \quad , \tag{1}$$

where I_0 is the on-axis peak intensity and $L_{eff} = \dfrac{1 - e^{-\alpha d}}{\alpha}$ is the effective optical length of the sample with thickness *d* and linear absorption coefficient α. *z* is the sample position relative to the beam focal point, and $z_0 = 2\pi\omega_0^2/\lambda$ is the confocal parameter of the beam with waist ω_0 and wavelength λ.

For the closed aperture, the transmittance is given by:

$$T_{close}(z) = 1 + \frac{4(\frac{z}{z_0})\Delta\Phi}{(1 + (\frac{z}{z_0})^2)(9 + (\frac{z}{z_0})^2)} - \frac{\beta I_0 L_{eff}}{2\sqrt{2}} \frac{(3 - (\frac{z}{z_0})^2)}{(1 + (\frac{z}{z_0})^2)(9 + (\frac{z}{z_0})^2)} \quad ; \tag{2}$$

here, $\Delta\Phi$ is the time-averaged phase change induced by the nonlinear refraction, which can be approximated as:

$$\Delta\Phi = \frac{2\pi}{\sqrt{2}\lambda}(1 - S)^{0.25} n_2 I_0 L_{eff} \quad , \tag{3}$$

where *S* is the transmissivity of the aperture.

The Gaussian decomposition method requires the proper determination of the effective optical length, L_{eff}, which can be deduced from the linear optical constants of the *c*-Si and *p*-Si samples. For the *c*-Si samples, the linear absorption coefficient can be estimated by $\alpha = \dfrac{4\pi k}{\lambda}$, where *k* is the imaginary part of linear refractive index. At an 800-nm wavelength, $k = 0.0065$, as it was estimated by the spectroscopic ellipsometry method by Aspnes et al. [19], corresponding to $\alpha = 1.03 \times 10^3$ cm^{-1}. However, the refractive index and absorption coefficient values for *p*-Si are less known since these values depend on the porosity, silicon skeleton dimensionality and surface adsorbates [20]. It has been shown previously that the Maxwell–Garnett mixing rules can be applied in order to calculate the linear complex permittivity of *p*-Si at 800 nm [21,22]:

$$\epsilon_{p-Si} = \epsilon_{c-Si} + 2\epsilon_{c-Si} \frac{f \dfrac{\epsilon_{air} - \epsilon_{c-Si}}{\epsilon_{air} + \epsilon_{c-Si}}}{1 - f \dfrac{\epsilon_{air} - \epsilon_{c-Si}}{\epsilon_{air} + \epsilon_{c-Si}}} \tag{4}$$

where ϵ_{c-Si} and ϵ_{air} are the permittivity of *c*-Si and air pores ($\epsilon_{air} = 1$), respectively. f is the volume fraction of air pores, which was estimated to be 68% for the samples used in this work. Thus, we obtain that $k = 0.0025$ and $\alpha = 386$ cm^{-1}. It is worth mentioning that the Maxwell–Garnett mixing rules are valid for these samples since the wavelength is about two orders of magnitude greater than the dimensions of the *p*-Si components.

As for the 2.4-μm wavelength, there were no explicitly known values for the imaginary part of the refractive index of intrinsic *c*-Si. This was due to the insignificant absorption for the incident photon energy much smaller than the band-gap. However, the silicon skeleton of *p*-Si that was used in this work was boron-doped, which provided about 3×10^{18} cm^{-3} free carriers [23]. Hence, the free carrier absorption needed to be considered, which further led to an increase of the k value. The influence of free carriers introduced by boron dopants on the *c*-Si dielectric function can be estimated using the Drude model:

$$\epsilon_{doped} = \epsilon_{undoped} - \frac{\dfrac{Ne^2}{m_h \epsilon_0}}{(\omega^2 + i\gamma\omega)}; \tag{5}$$

where N is the carrier concentration induced by dopants, e denotes the electron charge, $m_h = 0.36$ is the effective hole mass and ϵ_0 is the vacuum permittivity. The carrier-carrier scattering rate γ was chosen to be $1 \times 10^{13} (s^{-1})$, which is a typical value for doped silicon [24]. Further use of the Maxwell–Garnett model for the *p*-Si provided the value of the dielectric function of $\epsilon_{p-Si} = 3.55 + 0.009i$. The corresponding absorption coefficient, α, was 1.54 cm^{-1} at 2.4 μm. We note that the dopant-induced free carrier absorption was about two orders of magnitude lower than the intrinsic absorption of silicon at 800 nm. Therefore, it could be safely discarded in the calculations for an 800-nm wavelength.

Figure 1a,b shows the measurements taken with the closed and open aperture, respectively, of Z-scan transmittance of the 9 μm-thick *c*-Si thin film, using incident fluences of 40, 54, 68 and 82 nJ. The experimental results were fitted with Equations (1) and (2) to extract the TPA coefficient, β, and nonlinear refractive index, n_2. The fitting results suggest that $\beta = 2.9 \pm 0.9$ cm/GW, while $n_2 = 7 \pm 2 \times 10^{-6}$ cm^2/GW, which are comparable to those previously published elsewhere [16,17]. The main reason for the observed uncertainties was due to the experimental error of the beam fluence determination. The fluence of the incident light did not exceed the nano-Joule range, and it was difficult to measure its exact value. The error involved in the measurements of the fluence was up to 30%. However, this should not lead to a severe numerical error and deviation from the acceptable standards for these types of experiments.

The experimental procedure was repeated to measure the non-linear response of a 15 μm-thick free-standing *p*-Si thin film, and the results are shown in Figure 2. We note, the incident fluence covered a slightly different range than that used for the *c*-Si's measurements. This was due to the difficulties in controlling the fluence level in such a low pulse energy range. The analysis revealed the TPA coefficient $\beta = 1.0 \pm 0.3$ cm/GW and nonlinear refractive index $n_2 = -9 \pm 3 \times 10^{-5}$ cm^2/GW. In comparison with that of *c*-Si, the TPA appeared to be slightly suppressed. However, the nonlinear refraction was greatly enhanced. Remarkably, the sign of n_2 was reversed to the negative in comparison to the positive value of *c*-Si. These findings agree with Lettieri's work [25] on *p*-Si, despite the fact that we used samples with different doping and porosity.

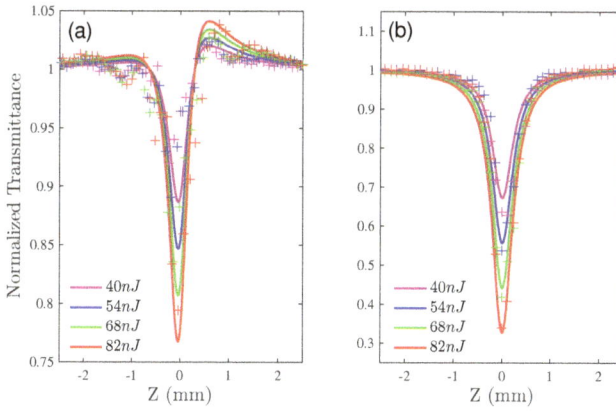

Figure 1. The experimental (dots) and fitting (solid lines) results of the Z-scan measurements for the 9 μm-thick-*c*-Si sample at 800 nm, using incident fluence of 40 (magenta), 54 (blue), 68 (green) and 82 nJ (red), respectively. (**a**) Closed aperture Z-scan (normalised transmittance with a 500-μm aperture). (**b**) Open aperture Z-scan (normalised transmittance without aperture).

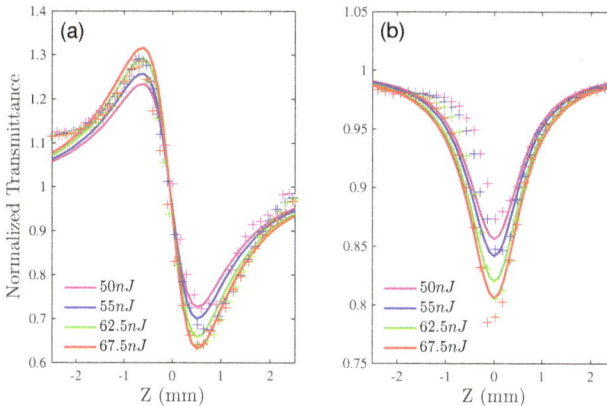

Figure 2. The experimental (dots) and fitting (solid lines) results of the Z-scan measurements for the 15 μm-thick *p*-Si sample at 800 nm, using incident fluence of 50 (magenta), 55 (blue), 62.5 (green) and 67.5 nJ (red), respectively. (**a**) Closed aperture Z-scan (normalised transmittance with a 500-μm aperture). (**b**) Open aperture Z-scan (normalised transmittance without aperture).

To further compare the nonlinear refraction process between *c*-Si and *p*-Si, the nonlinear phase shift $\Delta\Phi$ for different incident fluences is plotted in Figure 3. The squares represent the experimental values $\Delta\Phi = \Delta T_{peak-valley}$, and the solid lines are calculated from Equation (3) with the values of n_2 obtained from the Z-scan measurements. It can be seen that the phase shift for *p*-Si was at least four-times higher and changed faster as a function of the fluence than that of the *c*-Si, which further confirmed the enhancement of the nonlinear refraction.

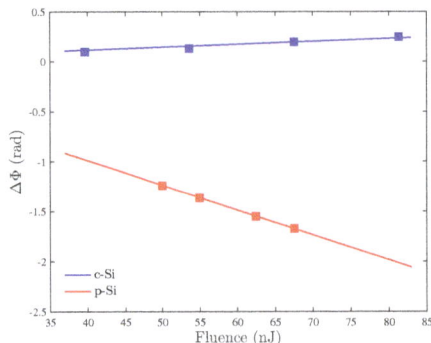

Figure 3. Nonlinear phase shift, $\Delta\Phi$, as a function of the incident fluence for *c*-Si (blue) and *p*-Si (red). The squares represent the experimental results, and the solid lines are calculated with n_2 obtained from the fitting results.

To further extend our knowledge about *p*-Si nonlinearity, we performed the experiments in the Short Wavelength Infrared (SWIR) where the photon energy is lower than the band-gap. Figure 4 shows the Z-scan results of *p*-Si at the wavelength of 2.4 μm. Because of the detectors' poor sensitivity at this wavelength range, the fluence was increased by an order of magnitude. The use of the longer wavelength required an increase in the scanning range from 5 mm to 30 mm to capture the stretched Z-scan features for both the closed and the open aperture cases. Due to the low level of the TPA at 2.4 μm, the open aperture experiments, as shown in Figure 4b, encountered a high noise level and problems preventing achieving the optimal alignment. The combination of these factors led to a distortion of the symmetrical shape about zero and a higher absolute error in the determination of the TPA coefficient. By fitting the experimental results, we obtained the nonlinear coefficients $\beta = 5.0 \pm 0.5 \times 10^{-3}$ cm/GW and $n_2 = -3.5 \pm 0.4 \times 10^{-4}$ cm^2/GW, indicating that the nonlinear refraction was even stronger for the longer wavelength range and that nonlinear absorption process was further restrained.

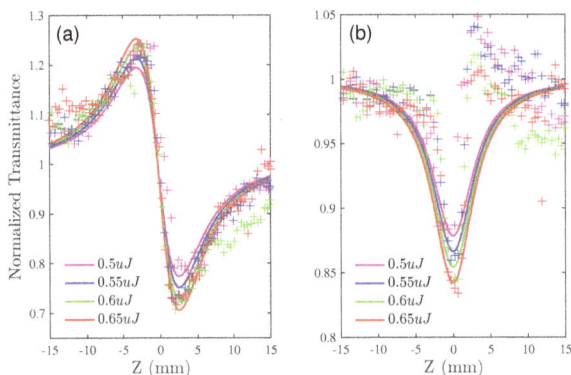

Figure 4. The experimental (dots) and fitting (solid lines) results of the Z-scan measurements for the 15 μm-thick *p*-Si sample for a pulse with wavelength of 2.4 μm and the incident fluence of 0.5 (magenta), 0.55 (blue), 0.6 (green) and 0.65 μJ (red), respectively. (**a**) Closed aperture Z-scan (normalised transmittance with a 500-μm aperture). (**b**) Open aperture Z-scan (normalised transmittance without aperture).

3. Discussion

The porous silicon used in this work consists of quasi-one-dimensional wires with a mean diameter of 10 nm. Therefore, the motion of the charged carriers is moderately confined. As a result of the confinement effect, the optical matrix element, defining the third order-nonlinear susceptibility (χ^3), increases. S. Lettieri et al. [25] proposed a computational model to calculate χ^3 as a function of quantum wire radius. It was suggested that for a quantum wire with a 2–3.5-nm radius, the real part of χ^3 is about four orders of magnitude greater than in the bulk. From our experimental results, we obtained $\Re(\chi^3) = \frac{4}{3}n_2 n_0^2 \epsilon_0 c = 1.2 \times 10^{-19}\ \mathrm{m^2/V^2}$ for *p*-Si, which is an order of magnitude greater than $3.37 \times 10^{-20}\ \mathrm{m^2/V^2}$ for *c*-Si. The difference of our result as compared to Lettieri's work might be the weaker confinement in our samples.

Moreover, in addition to the confinement, the optical Stark effect plays an important role in low-dimensional systems and can cause the sign reversal of n_2 (proportional to $\Re(\chi^3)$) in *p*-Si. To demonstrate this effect, D. Cotter et al. employed the sum-over-states method for the calculation of χ^3 of a sphere in a quantum confinement regime as a function of the sphere radius [14]. The results demonstrated that $\Re(\chi^3)$ is always negative with positive values of $\Im(\chi^3)$. This behaviour is different from the bulk materials, which exhibit negative values $\Re(\chi^3)$ only for photon frequencies very close to the absorption edge.

Our experimental results also show a suppression of the TPA process for an 800-nm photon wavelength and smaller β in *p*-Si with respect to *c*-Si. However, it should be noted that the TPA coefficient we evaluated in this work is an effective value of *p*-Si. Due to the dilution of silicon with air, the effective β is smaller than that of the silicon constituent, despite possible enhancement by the quantum confinement effect.

In summary, *p*-Si is a nonlinear material where the nonlinear refraction is greatly enhanced, but the nonlinear absorption is slightly suppressed with respect to its bulk counterpart. Such combinations of the coefficients make *p*-Si a suitable material for all-optical switching. The requirement for all optical switching is $|\frac{n_2}{\beta\lambda}| > c_{sw}$, where λ is the wavelength and c_{sw} is a switching parameter [26], the value of which depends on a particular switching scheme. For example, for a nonlinear coupled waveguide, $c_{sw} = 2$. Our results suggest that for 800-nm *p*-Si and *c*-Si, $|\frac{n_2}{\beta\lambda}| = 1.19$ and 0.03, respectively. However, at 2.4 µm, *p*-Si gives a promising value for all-optical switching application since $|\frac{n_2}{\beta\lambda}| = 292$. In comparison to other semiconductor composites that have been proposed for all-optical switching (for example Al-Ga-As [27], Ge-As-Se and Ge-As-S-Se demonstrated $|\frac{n_2}{\beta\lambda}| > 5$ [28]), *p*-Si has the advantage of the established fabrication technology, relatively low cost and less contamination to the environment.

4. Materials and Methods

4.1. Femtosecond Laser System

The experiments in this work were carried out using a Coherent femtosecond laser system. The seed laser was generated from a mode-locked oscillator and further amplified in a Q-switched amplification cavity. The output laser beam has a spatial Gaussian distribution, and the central wavelength was around 800 nm with a 100-nm spectral width. The pulse duration was about 60 fs with a 1-kHz repetition rate. More information on the laser system can be found elsewhere [6,22,29]. In addition, an optical parametric amplifier (OPerA-Solo) was used for the generation of a 2.4-µm short infrared laser. The samples were placed on a motorised linear stage (VT-80) and translated through the focus of the Gaussian beam with a step size of 0.15 mm. The beam was focused to a minimum beam waist of 16 µm by an N-BK7 Plano-convex lens with a 60-mm focal length. The pulse fluence was controlled by a half-wave plate combined with a linear polariser, which sets the polarisation to the in-plane direction (p-polarised). All experiments were conducted at the normal incidence.

We note that in general, *p*-Si can be classified as a uniaxial birefringent material with the optical axis aligned along the pores' direction [30–32]. However, at the normal incidence, there is no electric field component parallel to that axis, and the material can be treated as ordinary. A 45%/55% beam splitter was placed behind the sample, which splits the transmitted signal to two identical detectors, one with a 500-μm aperture ($S \approx 0.05$) in front and one without for the closed and open aperture measurements, respectively. Silicon detectors with a 5×5 mm detection area were used for the 800-nm Z-scan and PbS detectors for the 2.4-μm Z-scan.

4.2. Sample Preparation

Electro-chemical anodization was used for the preparation of *p*-Si samples. A boron-doped Si (100) wafer (resistivity: 5 to 15 mΩcm) was immersed in a mixture of methanol and 40% hydrofluoric acid electrolyte (mixing ratio 1:1). A current density of 30 mA/cm^2 was then applied for the anodization process. The *p*-Si layer was separated from the underlying substrate by applying a short pulsed current with the current density of 120 mA/cm^2. The membrane was then rinsed in methanol and dried in air. A porosity of 68% was measured using gravimetric analysis, and the thickness of 15 μm was estimated from the images of the cross-section obtained by a scanning electron microscope. The structure of the obtained samples can be described as sponge-like silicon material consisting of interwoven wires with a mean diameter of about 10 nm. More details about the sample characterisation can be found in [23,33].

5. Conclusions

In conclusion, we reported an enhancement on nonlinear refraction and suppression on TPA of *p*-Si in comparison with *c*-Si at 800 nm. The nonlinear refractive indices that we obtained for *c*-Si and *p*-Si are $7 \pm 2 \times 10^{-6}$ cm^2/GW and $-9 \pm 3 \times 10^{-5}$ cm^2/GW, respectively. The TPA coefficient was determined to be $\beta = 2.9 \pm 0.9$ cm/GW for *c*-Si and $\beta = 1.0 \pm 0.3$ cm/GW for *p*-Si. In addition, a further increase of n_2 and a decrease of β were found with an incident wavelength of 2.4 μm. We believe that the quantum confinement in *p*-Si is the main reason for the above findings.

Author Contributions: Conceptualization, A.K.; Data curation, R.W.; Formal analysis, R.W.; Methodology, L.C.; Software, J.C.; Supervision, A.K.; Writing—original draft, R.W.; Writing—review & editing, J.C., .C. and A.K.

Funding: This research received no external funding.

Acknowledgments: We would like to thank Emmalene Wilson for her contributions on developing the Z-scan method and Dimitri Chekulaev for his help with the laser system.

Conflicts of Interest: The authors declare no conflict of interest.

References

1. Lehmann, V.; Gosele, U. Porous silicon formation: A quantum wire effect. *Appl. Phys. Lett.* **1991**, *58*, 856–858. [CrossRef]
2. Zakar, A.; Park, S.J.; Zerova, V.; Kaplan, A.; Canham, L.T.; Lewis, K.L.; Burgess, C.D. MWIR optical modulation using structured silicon membranes. *Int. Soc. Opt. Photonics* **2016**, *9992*, 999203. [CrossRef]
3. Cullis, A.G.; Canham, L.T.; Calcott, P.D.J. The structural and luminescence properties of porous silicon. *J. Appl. Phys.* **1997**, *82.3*, 909–965. [CrossRef]
4. Klimov, V.; McBranch, D.; Karavanskii, V. Strong optical nonlinearities in porous silicon: Femtosecond nonlinear transmission study. *Phys. Rev. B* **1995**, *52*, R16989. [CrossRef]
5. Henari, F.Z.; Morgenstern, K.; Blau, W.J.; Karavanskii, V.A.; Dneprovskii, V.S. Third order optical nonlinearity and all-optical switching in porous silicon. *Appl. Phys. Lett.* **1995**, *67*, 323–325. [CrossRef]
6. Park, S.J.; Zakar, A.; Zerova, V.L.; Chekulaev, D.; Canham, L.T.; Kaplan, A. All-optical modulation in Mid-Wavelength Infrared using porous Si membranes. *Sci. Rep.* **2016**, *6*, 30211. [CrossRef] [PubMed]
7. Lin, V.S.Y.; Motesharei, K.; Dancil, K.P.S.; Sailor, M.J.; Ghadiri, M.R. A porous silicon-based optical interferometric biosensor. *Science* **1997**, *278*, 840–843. [CrossRef] [PubMed]

8. Dancil, K.P.S.; Greiner, D.P.; Sailor, M.J. A porous silicon optical biosensor: Detection of reversible binding of IgG to a protein A-modified surface. *J. Am. Chem. Soc.* **1999**, *121*, 7925–7930. [CrossRef]

9. Menna, P.; Di Francia, G.; La Ferrara, V. Porous silicon in solar cells: A review and a description of its application as an AR coating. *Sol. Energy Mater. Sol. Cells* **1995**, *37*, 13–24. [CrossRef]

10. Li, Y.Y.; Cunin, F.; Link, J.R.; Gao, T.; Betts, R.E.; Reiver, S.H.; Chin, V.; Bhatia, S.N.; Sailor, M.J. Polymer replicas of photonic porous silicon for sensing and drug delivery applications. *Science* **2003**, *299*, 2045–2047. [CrossRef] [PubMed]

11. Qiu, J. Femtosecond laser-induced microstructures in glasses and applications in micro-optics. *Chem. Rec.* **2004**, *4*, 50–58. [CrossRef] [PubMed]

12. Ashkenasi, D.; Varel, H.; Rosenfeld, A.; Henz, S.; Herrmann, J.; Cambell, E.E.B. Application of self-focusing of ps laser pulses for three-dimensional microstructuring of transparent materials. *Appl. Phys. Lett.* **1998**, *72*, 1442–1444. [CrossRef]

13. Sheik-Bahae, M.; Said, A.A.; Wei, T.H.; Hagan, D.J.; Van Stryland, E.W. Sensitive measurement of optical nonlinearities using a single beam. *IEEE J. Quantum Electron.* **1990**, *26*, 760–769. [CrossRef]

14. Cotter, D.; Burt, M.G.; Manning, R.J. Below-band-gap third-order optical nonlinearity of nanometer-size semiconductor crystallites. *Phys. Rev. Lett.* **1992**, *68*, 1200. [CrossRef] [PubMed]

15. Weaire, D.; Wherrett, B.S.; Miller, D.A.B.; Smith, S.D. Effect of low-power nonlinear refraction on laser-beam propagation in InSb. *Opt. Lett.* **1979**, *4*, 331–333. [CrossRef] [PubMed]

16. Dinu, M.; Quochi, F.; Garcia, H. Third-order nonlinearities in silicon at telecom wavelengths. *Appl. Phys. Lett.* **2003**, *82*, 2954–2956. [CrossRef]

17. Bristow, A.D.; Rotenberg, N.; Van Driel, H.M. TPA and Kerr coefficients of silicon for 850–2200 nm. *Appl. Phys. Lett.* **2007**, *90*, 191104. [CrossRef]

18. Gu, B.; Chen, J.; Fan, Y.X.; Ding, J.; Wang, H.T. Theory of Gaussian beam Z scan with simultaneous third-and fifth-order nonlinear refraction based on a Gaussian decomposition method. *JOSA B* **2005**, *22*, 2651–2659. [CrossRef]

19. Aspnes, D.E.; Studna, A.A. Dielectric functions and optical parameters of Si, Ge, GaP, GaAs, GaSb, InP, InAs, and InSb from 1.5 to 6.0 ev. *Phys. Rev. B* **1983**, *27*, 985. [CrossRef]

20. Canham, L. (Ed.) *Handbook on Porous Silicon*; Springer: Basel, Switzerland, 2014; ISBN 978-3-319-05744-6.

21. Sihvola, A.H. *Electromagnetic Mixing Formulas and Applications*, Clarricoats, P.J.B., Jull, E.V., Eds.; The Institute of ElectricalEngineers: London, UK, 1999; ISBN 9780852967720.

22. Chekulaev, D.; Garber, V.; Kaplan, A. Free carrier plasma optical response and dynamics in strongly pumped silicon nanopillars. *J. Appl. Phys.* **2013**, *113*, 143101. [CrossRef]

23. Zakar, A.; Wu, R.; Chekulaev, D.; Zerova, V.; He, W.; Canham, L.; Kaplan, A. Carrier dynamics and surface vibration-assisted Auger recombination in porous silicon. *Phys. Rev. B* **2018**, *97*. [CrossRef]

24. Qiu, B.; Tian, Z.; Vallabhaneni, A.; Liao, B.; Mendoza, J.M.; Restrepo, O.D.; Ruan, X.; Chen, G. First-principles simulation of electron mean-free-path spectra and thermoelectric properties in silicon. *Europhys. Lett.* **2015**, *109*, 57006. [CrossRef]

25. Lettieri, S.; Fiore, O.; Maddalena, P.; Ninno, D.; Di Francia, G.; La Ferrara, V. Nonlinear optical refraction of free-standing porous silicon layers. *Opt. Commun.* **1999**, *168*, 383–391. [CrossRef]

26. Sheik-Bahae, M.; Hutchings, D.C.; Hagan, D.J.; Van Stryland, E.W. Dispersion of bound electron nonlinear refraction in solids. *IEEE J. Quantum Electron.* **1991**, *27*, 1296–1309. [CrossRef]

27. Siviloglou, G.A.; Suntsov, S.; El-Ganainy, R.; Iwanow, R.; Stegeman, G.I.; Christodoulides, D.N.; Pozzi, F. Enhanced third-order nonlinear effects in optical AlGaAs nanowires. *Opt. Express* **2006**, *14*, 9377–9384. [CrossRef] [PubMed]

28. Harbold, J.M.; Ilday, F.O.; Wise, F.W.; Aitken, B.G. Highly nonlinear Ge-As-Se and Ge-As-S-Se glasses for all-optical switching. *IEEE Photonics Technol. Lett.* **2002**, *14*, 822–824. [CrossRef]

29. Roger, T.W.; He, W.; Yurkevich, I.V.; Kaplan, A. Enhanced carrier-carrier interaction in optically pumped hydrogenated nanocrystalline silicon. *Appl. Phys. Lett.* **2012**, *101*, 141904. [CrossRef]

30. Cazzanelli, M.; Kovalev, D.; Dal Negro, L.; Gaburro, Z.; Pavesi, L. Polarized optical gain and polarization-narrowing of heavily oxidized porous silicon. *Phys. Rev. Lett.* **2004**, *93*, 207402. [CrossRef] [PubMed]

31. Kunzner, N.; Diener, J.; Gross, E.; Kovalev, D.; Timoshenko, V.Y.; Fujii, M. Form birefringence of anisotropically nanostructured silicon. *Phys. Rev. B* **2005**, *71*, 195304. [CrossRef]

32. Golovan, L.A.; Timoshenko, V.Y.; Fedotov, A.B.; Kuznetsova, L.P.; Sidorov-Biryukov, D.A.; Kashkarov, P.K.; Diener, J. Phase matching of second-harmonic generation in birefringent porous silicon. *Appl. Phys. B* **2001**, *73*, 31–34. [CrossRef]

33. He, W.; Yurkevich, I.V.; Canham, L.T.; Loni, A.; Kaplan, A. Determination of excitation profile and dielectric function spatial nonuniformity in porous silicon by using WKB approach. *Opt. Express* **2014**, *22*, 27123. [CrossRef] [PubMed]

© 2018 by the authors. Licensee MDPI, Basel, Switzerland. This article is an open access article distributed under the terms and conditions of the Creative Commons Attribution (CC BY) license (http://creativecommons.org/licenses/by/4.0/).

MDPI
St. Alban-Anlage 66
4052 Basel
Switzerland
Tel. +41 61 683 77 34
Fax +41 61 302 89 18
www.mdpi.com

Applied Sciences Editorial Office
E-mail: applsci@mdpi.com
www.mdpi.com/journal/applsci

www.ingramcontent.com/pod-product-compliance
Lightning Source LLC
Chambersburg PA
CBHW041217220326
41597CB00033BA/5993